REVISE KEY STAGE 3
Mathematics
STUDY GUIDE
Foundation

Series Consultant: Harry Smith

Authors: Bobbie Johns and Sharon Bolger

Our study resources are the smart choice for those studying Key Stage 3 Mathematics, and preparing to start the GCSE (9-1) Mathematics Foundation course. This Study Guide includes a FREE online edition and will help you to:

- **Organise** your study with the one-topic-per-page format
- **Speed up** your study with helpful hints
- **Track** your progress with at-a-glance check boxes
- **Check** your understanding with Worked Examples
- **Develop** your technique with practice questions and full answers
- **Progress** towards the GCSE (9-1) Maths Foundation course with problem-solving practice.

Check out the Key Stage 3 Study Workbook too!

Make sure that you have practised every topic covered in this book, with the accompanying Key Stage 3 Mathematics Study Workbook. It gives you:

- More practice questions and a 1-to-1 page match with this Study Guide
- Guided questions to help build your confidence
- Hints to support your study and practice.

For the full range of Pearson revision titles across KS3, GCSE, AS/A Level and BTEC visit:
www.pearsonschools.co.uk/revise

Contents

Whole numbers

When you compare and order **whole numbers**, remember that the more digits a number has, the larger it is. When two numbers have the same number of digits, compare the digits, one by one, starting from the left.

3497 and 3502 both have **4 digits**. They both have **3 thousands**. 3497 has **4 hundreds** but 3502 has **5 hundreds**, so 3502 is larger.

Worked example

(a) Write the number 34 507 in words.

Thirty-four thousand, five hundred and seven

(b) Write the number 1 320 045 in words.

One million, three hundred and twenty thousand and forty-five

(c) Write five hundred and nine thousand and four in figures.

509 004

Write the numbers in a place value diagram:

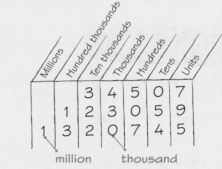

Worked example

(a) Write the value of the digit 4 in
    (i) 2411   (ii) 43 850   (iii) 994   (iv) 4 987 653
(i) 400   (ii) 40 000   (iii) 4   (iv) 4 000 000

(b) Write these numbers in order of size.
    Start with the **smallest** number.
    2408   2954   43 850   2411   944

944   2408   2411   2954   43 850

(c) Write these numbers in order of size.
    Start with the **largest** number.
    375 890   380 001   2 000 000   99 999

7 digit > 6 digit and 380 000 > 375 000
2 000 000   380 001   375 890   99 999

The value of a digit depends on its position in the number.

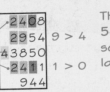

The thousands digits are the same, so compare the hundreds.

9 > 4
1 > 0

There is only 5-digit number, so it is the largest.

There is only one 3-digit number, so it is the smallest.

Negative numbers

Numbers less than zero are negative numbers. On this number line, the negative numbers are to the left of 0. The further the number is to the left, the less its value.

−10 −5   0   5   10   −10 is less than −5

Worked example

Write these numbers in order of size.
Start with the **smallest** number.
−3   6   0   −9   10   −2   4
−9   −3   −2   0   4   6   10

Now try this

1 The population of a town is 95 364.
  (a) Write the number 95 364 in words.
  (b) What is the value of the digit 5 in 95 364?

2 Write these numbers in order of size. Start with the **smallest** number.
  (a) 21 089   20 098   1 000 010   3756   21 465   3765
  (b) −3   6   −11   4   9   −1

3 Look at the thermometers. Which city has the lower temperature?

Moscow

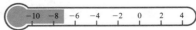

Helsinki

Decimals

To compare and order decimal numbers, you need to understand the place value of each digit.

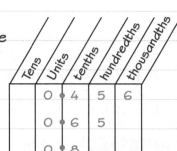

0.456 ◀——— this number has 4 tenths, 5 hundredths and 6 thousandths

0.65 ◀——— this number also has 5 hundredths, but it has 6 tenths

0.8 ◀——— this number has 8 tenths

Worked example

(a) Write the value of the digit 5 in each decimal number.
   (i) 0.625   (ii) 0.562   (iii) 0.652

(i) 5 thousandths   (ii) 5 tenths
(iii) 5 hundredths

(b) Write these decimals in order of size, smallest first.
   0.45   0.7   0.445   0.54

4 < 7 so 0.45 < 0.7
0.445 < 0.45 as it has fewer hundredths
0.7 > 0.54 because 7 tenths > 5 tenths
0.445   0.45   0.54   0.7

> The first decimal place is tenths, the second is hundredths and the third is thousandths.

> You can write in zeros to make the decimals all the same length. Then you can use your knowledge of whole numbers to write them in order:
>
> 0.45**0**   0.7**00**   0.445   0.54**0**

Decimals and measurements

You often use decimals when you write measurements.

0.5 cm   3.8 cm   8.2 cm   13.7 cm

Worked example

Write the measurements shown by the arrows in metres.   (a)   (b)   (c)   (d)

(a) 1.28 m   (b) 1.31 m   (c) 1.345 m
(d) 1.375 m

(a) Count the large markers up from 1.25: 1.25, 1.26, 1.27, **1.28**

(b) 1.3 = 1.30. The next large marker is 1.31

(c) The number halfway between 1.34 and 1.35 is 1.345

(d) The number halfway between 1.37 and 1.38 is 1.375

Now try this

1 Use < or > to compare the decimals in each pair.

(a) 6.7 and 6.456   (b) 23.819 and 23.84

2 Write these decimals in order of size, smallest first.   2.3   2.15   2.199   2.26

3 What is the value of the digit 6 in 23.067?

4 Whose suitcase is heavier?

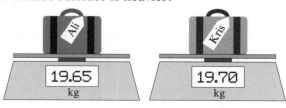

19.65 kg   19.70 kg

Rounding

Rounding a number gives an acceptable approximation.

To round a number look at the next digit to the right on a place value diagram.

5 or more → round up, less than 5 → round down.

You can round a decimal number to the nearest whole number or to a given number of decimal places (d.p.).

To round to the nearest whole number look at the digit in the tenths column.
It is a 5, so round up.
6.504 rounded to the nearest whole number is 7.

Units	.	tenths	hundredths	thousandths
6	•	5	0	4

To round to 1 d.p. look at the digit in the second decimal place.
It is 0, so round down.
6.504 rounded to 1 d.p. is 6.5.

To round to 2 d.p. look at the digit in the third decimal place.
It is 4, so round down.
6.504 rounded to 2 d.p. is 6.50.
You **need** to write the 0 to show you have rounded to 2 d.p.

Worked example

Round 4.574 91 to
(a) the nearest whole number   (b) 1 d.p.
(c) 2 d.p.

(a) 5   (b) 4.6   (c) 4.57

Look at the next digit to the right.
(a) 4.57491 → 5, so round up
(b) 4.57491 → 7, so round up
(c) 4.57491 → 4, so round down

Worked example

Use rounding to estimate the answers.
(a) 3912 − 1937
4000 − 2000 = 2000
(b) 38 × 42
40 × 30 = 1200
(c) 4.6 × 9.8
5 × 10 = 50

- To **estimate** an answer, round each number to **1 significant figure**. Then work out the calculation.
- You can use mental strategies to speed up your calculations:
  40 × 30 = 4 × 3 × 10 × 10
              = 12 × 100 = 1200

Significant figures

Rounding numbers in calculations to 1 s.f. is a useful way to estimate the answer. Always start counting significant figures (s.f.) from the **left**.

27.05 rounded to 1 s.f. is 30

Look for the digit furthest to the left, which is 2. The next digit is 7, so round up to give an answer of 30.

27.05 rounded to 2 s.f. is 27.

Look for the two digits furthest to the left, which are 2 and 7. The next digit is 0, so round down to give an answer of 27.

You start counting significant figures with the first **non-zero** digit. In 0.291 the first significant figure is 2 tenths. In 0.685 the first significant figure is 6 tenths, and the second significant figure is 8 hundredths.

Worked example

Round these numbers to the given number of significant figures (s.f.).
(a) 2671 to 1 s.f.   (b) 45 672 to 3 s.f.
3000   45 700
(c) 0.291 to 1 s.f.   (d) 0.685 to 2 s.f.
0.3   0.69

Now try this

1 Lara says that 3790 + 2858 is more than 7000. Use rounding to show that she is incorrect.

2 Use rounding to estimate the answer to 5.7 × 3.2

3 Round these numbers to 2 s.f.
(a) 3829   (b) 24.93   (c) 0.7354

Addition

You can use mental or written methods to add whole numbers and decimal numbers.

Column addition

319 + 76 + 2045

Write the numbers as a vertical list, in size order, making sure the digits are in the correct columns.

```
        Th  H  T  U
         2  0  4  5
319 + 76 + 2045 = 2440
            3  1  9
               7  6
        _____
         2  4  4  0
            1  2
```

Add the units first: 5 + 9 + 6 = 20
Write 0 in the answer line in the units column, and write the 2 under the tens column.

Add the tens, remembering to add the carried 2 (14). Write the 4 in the answer line and the 1 under the hundreds column.

Add the hundreds, writing the answer in the answer. Then write the thousands directly in the answer line.

Find the total of
(a) 72.65 and 49.937   (b) 0.785, 1.96 and 17.5

```
 H  T  U . t  h  th
 7  2  6 . 6  5  0
 4  9  9 . 9  3  7
_____
1  2  2 . 5  8  7
    1  1
```

```
    T  U . t  h  th
    1  7 . 5  0  0
     1 . 9  6  0
     0 . 7  8  5
_____
    2  0 . 2  4  5
       2  2  1
```

Here is Ellie's till receipt.
She has £50.
Is that enough money to pay
for all the items?

| £14.51 |
| £24.75 |
| £0.50 |
| £10.05 |

```
 2  4 . 7  5
 1  4 . 5  1
 1  0 . 0  5
    0 . 5  0
_____
 4  9 . 8  1
    1     1
```

£50 is enough.

Write the prices as a column addition.
Add the prices to work out the total.
Check if it is more or less than £50.

You'll need brilliant problem-solving skills
to succeed in GCSE – get practising now!

1 Work out 481 + 79 + 7904
2 Find the total of 25.9, 13.675 and 9.45
3 What is the perimeter of this rectangular field?

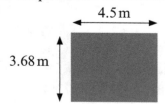

4.5 m

3.68 m

4 Who has spent more money? You must show
  your working.

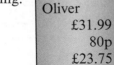

Oliver	Lukas
£31.99	£29.01
80p	75p
£23.75	£26.89

A rectangle has two pairs of equal sides.
Add all four sides to work out the perimeter.

Subtraction

You can use mental or written methods to subtract whole and decimal numbers.

A number line can be helpful, for example, to work out $124 - 32$:

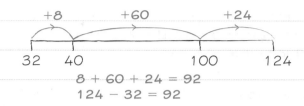

$8 + 60 + 24 = 92$
$124 - 32 = 92$

Column subtraction

$6492 - 3749$

Th	H	T	U
$^5\cancel{6}$	$^1\cancel{4}$	$^8\cancel{9}$	$^1 2$
3	7	4	9
2	7	4	3

- Write the calculation in vertical columns.
- You **must** subtract the bottom number from the top number.
- Start with the units: $2 < 9$ so take 1 ten from the tens column and change it into 10 units. Then you can subtract 9 units from 12 units.
- In the tens column, $8 > 4$ so simply subtract: $8 - 4$.

In the hundreds column, $4 < 7$ so change 1 thousand to 10 hundreds: $14 - 7$ hundreds. Subtract the remaining thousands.

Golden rules

✓ Line up digits with the same place value **and** the decimal points.
✓ Write the decimal point in the answer line.
✓ Write extra zeros to ensure that both numbers have the same number of decimal places.
✓ Subtract, column by column, starting from the right.

Worked example

Jamal	3.7 m
Kai	2.225 m

How much further did Jamal jump than Kai?

U	t	h	th
3 .	$^6\cancel{7}$	$^9\cancel{0}$	$^1 0$
2 .	2	2	5
1 .	4	7	5

1.475 m

Problem solved!

Write the numbers as a column subtraction with the larger number on top.
Write zeros so both numbers have three decimal places.
To make the subtraction easier, change 1 tenth in the top number into 10 hundredths (making 3.6¹00).
Then change one of these new hundredths into 10 thousandths (3.69¹0).
Now subtract, column by column, starting on the right.

> You'll need brilliant problem-solving skills to succeed in GCSE – get practising now!

Now try this

1 Work out $3856 - 2937$

2 How much heavier is 3.4 kg than 2.75 kg?   Write as a column subtraction.

3 Alex spends £32.74 and pays with two £20 notes. How much change does she get?

Understanding powers of 10

When you multiply by 10, 100 or 1000 the digits move to the left on a place value diagram.

$32 \times 100 = 3200 \qquad 1.25 \times 10 = 12.5$

When you divide by 10, 100 or 1000 the digits move to the right on a place value diagram.

$750 \div 100 = 7.5 \qquad 2780 \div 10 = 278$

Powers of 10

When 10 is raised to a whole-number power, the power is the same as the number of zeros.

$10^1 = 10 = 10$

$10^2 = 10 \times 10 = 100$

$10^3 = 10 \times 10 \times 10 = 1000$

Worked example

Work out
(a) $2.37 \times 100 \qquad = 237$
(b) $67.4 \div 100 \qquad = 0.674$
(c) $2856 \div 1000 \qquad = 2.856$
(d) $7825 \div 10^2 \qquad = 78.25$
(e) $83.24 \times 10^3 \qquad = 83\,240$

In part (b), when you divide by 100 the digits move **two places** to the **right**.
In part (d), $10^2 = 100$.

1 million $= 10^6 = 1\,000\,000$

To divide by $1\,000\,000$ move the digits 6 places to the right:

$$7\,074\,000 = 7.074 \text{ million}$$

To multiply by $1\,000\,000$ move the digits 6 places to the left:

$$0.616 \text{ million} = 616\,000$$

Worked example

Complete the table to write these populations as whole numbers and as millions.

City	Population	Population
London	7074000	7.074 million
Leeds	727000	0.727 million
Glasgow	616000	0.616 million

Negative powers

A power can also be negative.

$10^{-1} = 0.1 = \frac{1}{10} \qquad 10^{-2} = 0.01 = \frac{1}{100}$

See pages 12 and 14 for more on powers and indices.

Golden rules

$\times 0.1$ is the same as $\div 10$

$\times 0.01$ is the same as $\div 100$

$\div 0.1$ is the same as $\times 10$

$\div 0.01$ is the same as $\times 100$

Worked example

Work out
(a) $54 \times 10^{-1} \quad = 54 \times 0.1 = 54 \div 10 = 5.4$
(b) $78 \times 0.01 \quad = 78 \div 100 = 0.78$
(c) $67 \div 0.1 \quad = 67 \times 10 = 670$
(d) $35 \div 10^{-2} \quad = 35 \div 0.01 = 35 \times 100 = 3500$

(a) 54×0.1 is the same as 54 tenths.

(b) 78×0.01 is 78 hundredths.

(c) $67 \div 0.1$: how many tenths are in 67?

(d) $35 \div 0.01$: how many hundredths are in 35?

Now try this

1 Work out
(a) $618 \div 100$ (b) 5.78×10
(c) 83×0.1 (d) $37 \div 0.01$

2 Write 2×10^5 as a whole number.

3 A house is for sale for £450 000. Write the price as a decimal with the word 'million'.

Multiplication

You can use written methods to multiply whole numbers or decimals.

Worked example

(a) 2365 × 9

```
    2 3 6 5
  ×       9
  2 1 2 8 5
    3 5 4
```

(b) 438 × 24

```
        4 3 8
    ×     2 4
      1 7 5 2
      8 7 6 0
    1 0 5 1 2
        1 1
```

To multiply by a 2-digit number:
- multiply by the units
- multiply by the tens
- add these two answers.

Multiply each digit by 9, starting with the units.
Units: 9 × 5 = 45, so write 5 in the units column answer line and carry the 4 over to the tens column.
Tens: 9 × 6 = 54, + the carried 4
      = 58
Hundreds: 9 × 3 = 27, + the carried 5
      = 32
Thousands: 9 × 2 = 18, + the carried 3
      = 21

Multiplying decimals

You can use column multiplication to multiply decimals.

- Ignore the decimal point and work out the whole-number multiplication.
- Count the number of decimal places in the question.
- Put the same number of decimal places into the whole-number answer.

Worked example

Work out 2.<u>7</u> × 3.<u>9</u> (2d.p.)

Estimate: 3 × 4 = 12

```
        2 7
    ×   3 9
      2 4 3
      8 1 0
    1 0 5 3
```

27 × 39 = 1053
2.7 × 3.9 = 10.53

Use rounding to estimate.
Work out the whole-number multiplication.
There are two decimal places in the question so put two decimal places in the whole-number answer.

Now try this

1 Work out
   (a) 2856 × 7   (b) 718 × 36
   (c) 2.6 × 3.5   (d) 35.64 × 3.2

2 There are 24 packs of peas in a box.
   The mean number of peas in a pack is 355.
   How many peas are there in the box?

3 A phone call from a hotel phone costs 25p per minute.
   How much will a 48-minute call cost?

4 Ru says that the answer to 12.4 × 4.8 is 595.2
   Show or explain why he is not correct.

Work out 355 × 24.

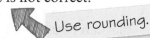
Use rounding.

Division of whole numbers

You need to be able to use mental and written methods to divide whole numbers. Knowing related multiplication facts and using multiples of 10 can help.

Worked example

Work out 276 ÷ 4

$$4)\overline{2\,^2 7\,^3 6}$$ ⁶ ⁹

276 ÷ 4 = 69

- 4 will not go into 2.
- 27 ÷ 4 = 6 remainder 3, so write 6 above the tens column and carry the 3 to the units.
- 36 ÷ 4 = 9. Write 9 in the units column.

When the divisor (the number you are dividing by) is bigger than 12:

- write out some multiples of the divisor
- divide in the normal way – some of the remainders might be more than 10
- 47 ÷ 18 = 2 remainder 11
- 115 ÷ 18 = 6 remainder 7

Worked example

Work out 4752 ÷ 18

Multiples of 18: 18, 36, 54, 72, 90, 108, …

$$18)\overline{4\,4 7\,^{11}5\,^7 2}$$ ² ⁶ ⁴

4752 ÷ 18 = 264

Worked example

702   284   725   501   450

Which of the numbers above are divisible by each of these numbers?

(a) 3   702, 501, 450
(b) 4   284
(c) 6   702, 450
(d) 9   702, 450
(e) 25   725, 450

Tests for divisibility

Use these facts to test whether a number is divisible by these whole numbers:

4   last two digits are a multiple of 4
5   last digit is 0 or 5
3   digit sum is a multiple of 3
6   even and digit sum is a multiple of 3
9   digit sum is a multiple of 9; repeated digit sum is 9
25   last 2 digits are 00, 25, 50 or 75

There might be more than one correct answer. As long as your answer works then it is correct.

Check using multiplication:

356 ÷ 4 = 89   89 × 4 = 356

Worked example

Arrange these digits so the division has no remainder:

☐4☐ ☐6☐ ☐5☐ ☐3☐

☐3☐ ☐5☐ ☐6☐ ÷ ☐4☐ = 89

The number must have 4 as its hundreds digit. The last two digits must be a multiple of 4 (so it must be even), and all three digits must add to a multiple of 9.

Now try this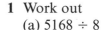

1 Work out
(a) 5168 ÷ 8   (b) 4224 ÷ 16

2 Eggs are packed in trays of 24.
How many trays are needed for 768 eggs?

3 Write a 3-digit number between 400 and 500 that can be divided by both 4 and 9.

Division with decimals

You need to be able to divide decimal numbers with and without a calculator. You can use the same methods as for whole numbers, but you need to take care with the decimal point.

Worked example

(a) A factory worker is packing jars into boxes. He has 7284 jars to pack. 24 jars will fit in a box. How many boxes will he need?

$7284 \div 24 = 303.5$ so 304 boxes

(b) Another worker is packing bottles into crates. Each crate holds 36 bottles. She has 7162 bottles. How many complete crates can she fill?

$7162 \div 36 = 198.94444...$ so 198 crates

Problem solved!

The answers must be **whole numbers**. Make sure you read the question carefully. In part (a), the worker needs to pack **all the jars**. 303 boxes is not enough, so round up to 304. For part (b), the worker needs to **fill** each crate. She can't fill 199 crates so round down to 198.

You'll need brilliant problem-solving skills to succeed in GCSE – get practising now!

Golden rules

✓ Make sure the decimal point in the answer line lines up with the decimal point in the number.

✓ Write extra zeros at the end of the decimal if necessary.

✓ To divide by a number with 1 decimal place, multiply **both numbers** by 10 first.

Worked example

(a) Work out $24.7 \div 5$

$$5 \overline{)2\,^24.\,^47\,^20} \quad 24.70 \div 5 = 4.94$$
answer: 4.94

(b) Work out $83.34 \div 1.8$

$83.34 \div 1.8 = 833.4 \div 18$

$$18 \overline{)83\,^{11}3.\,^54}$$
answer: 46.3

$83.34 \div 1.8 = 46.3$

Problem solved!

You need to know the multiples of 15 to divide by 15. Start by writing them out.
Remember to answer the question. You can work out £20 − £12.59 mentally by adding on in steps:

£12.59 → £13 → £20
  + £0.41   + £7

You'll need brilliant problem-solving skills to succeed in GCSE – get practising now!

Worked example

At a restaurant, 15 friends share the bill for £188.85 equally. Alison pays her share with a £20 note. How much change does she get?

Multiples of 15: 15, 30, 45, 60, 75, 90, 105, 120, 135, ...

$$15 \overline{)1\,^18\,^38.\,^88\,^{13}5}$$
answer: 12.59

Each person pays £12.59
£20 − £12.59 = £7.41 change

Now try this

1 Work out   (a) $735.2 \div 4$ (b) $696 \div 1.6$

2 Share £748.44 among 21 people.

3 How many 1.4 m lengths of fabric can be cut from a 21 m roll?

Negative numbers

Numbers that are less than zero are negative numbers. The + or − that tells you whether a number is positive or negative is called the **sign** of that number.

You need to know how to calculate with negative numbers.

Using number lines

You can use number lines to help you add or subtract with negative numbers.

See page 1 for more on ordering a set of negative numbers.

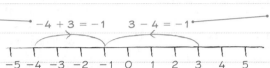

NEGATIVE NUMBERS | POSITIVE NUMBERS

−5 −4 −3 −2 −1 0 1 2 3 4 5

0 is neither positive nor negative.

When you add a positive number, move to the right on the number line.

$-4 + 3 = -1$ $3 - 4 = -1$

−5 −4 −3 −2 −1 0 1 2 3 4 5

When you subtract a positive number, move to the left on the number line.

Adding

Adding a negative number is the same as subtracting a positive number. The answer is **smaller** (further to the **left** on a number line).

$6 + -4 = 6 - 4 = 2$

Subtracting

Subtracting a negative number is the same as adding a positive number. The answer is **bigger** (further to the **right** on a number line).

$-1 - -9 = -1 + 9 = 8$

Golden rules

When **multiplying** or **dividing** two numbers:

- If the signs are **different** ⟶ the answer is **negative**.

- If the signs are the **same** ⟶ the answer is **positive**.

Worked example

Work out
(a) -5×8 $= -40$
(b) $36 \div -3$ $= -12$
(c) $-18 \div 2$ $= -9$
(d) -2×-5 $= 10$
(e) $-20 \div (-4) = 5$

Part (d) is negative × negative. The signs are the same so the answer is positive.

You can use a number line to work out additions and subtractions.

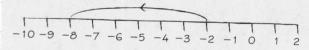

−10 −9 −8 −7 −6 −5 −4 −3 −2 −1 0 1 2

Worked example

One cold day in winter the daytime temperature in Alaska was –2°C. In the night it fell a further 6 degrees. What was the night-time temperature?

$-2 - 6 = -8$
The temperature was −8°C.

Now try this

Work out -2×-3 then multiply the result by -4.

1 Work out
   (a) $-6 + 7$ (b) $-9 - (-3)$
   (c) $-5 + (-4)$ (d) $-3 - 2$
2 Work out
   (a) -4×-3 (b) -9×4
   (c) $-24 \div (-4)$ (d) $-15 \div 3$

3 Max says $-2 \times -3 \times -4 = -24$, and Chris says it is 24. Who is right? Explain your answer.
4 The temperature in Stockholm in January was -4°C. The temperature in London was 6°C higher. What was the temperature in London?

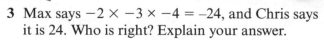
Use a number line.

Factors, multiples and primes

Understanding how to identify factors, multiples and primes is an important mathematical skill.

Factors

The factors of a number are whole numbers that divide into it exactly.

Factors of 18: 1, 2, 3, 6, 9, 18
Factors of 24: 1, 2, 3, 4, 6, 8, 12, 24

A **common factor** of two numbers is a factor of **both** numbers.

1, 2, 3 and 6 are common factors of 18 and 24.

The **highest common factor** (HCF) of two numbers is the largest of the common factors.

The HCF of 18 and 24 is 6.

Multiples

The multiples of a number are all the numbers in its times table.

Multiples of 6: 6, 12, 18, 24, 30, 36, 42, 48, ...
Multiples of 8: 8, 16, 24, 32, 40, 48, 56, ...

A **common multiple** of two numbers is a multiple of **both** numbers.

24 and 48 are common multiples of 6 and 8.

The **lowest common multiple** (LCM) of two numbers is the smallest of the common multiples.

The LCM of 6 and 8 is 24.

Worked example

Complete the diagram to show all the factors of 20.

The factors of 20 are 1, 2, 4, 5, 10 and 20

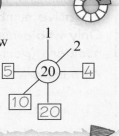

Factors come in **pairs**. The product of each **factor pair** is 20. On this diagram the factor pairs are **opposite** each other.

$1 \times 20 = 20$ $2 \times 10 = 20$ $4 \times 5 = 20$
1 and the number itself are always factors.

The LCM is the **smallest** number that appears in **both** lists.

Prime numbers

Prime numbers have exactly two factors: 1 and the number itself. There are an infinite number of prime numbers. You should learn the prime numbers smaller than 50:

2, 3, 5, 7, 11, 13, 17, 19, 23, 29, 31, 37, 41, 43, 47

2 is the smallest prime number, and is the only **even** prime number.
1 is **not** a prime number because it has only one factor – itself.

Worked example

(a) Write the first 10 multiples of 4.

4, 8, 12, 16, 20, 24, 28, 32, 36, 40

(b) Write the first five multiples of 9.

9, 18, 27, 36, 45

(c) Find the lowest common multiple of 4 and 9.

36

Now try this

1 Draw a diagram to show all the factors of 30.

2 (a) Find all the factors of (i) 32 (ii) 48
  (b) Write the HCF of 32 and 48.

3 Find the LCM of 14 and 8.

4 Write
  (a) a prime number that is between 20 and 30
  (b) a prime number that is 1 more than a multiple of 5.

Squares, cubes and roots

You should know:

- all the square numbers up to $10^2 = 100$ and their related square roots
- the first five cube numbers, as well as $10^3 = 1000$, and their related cube roots
- how to use the $\boxed{x^2}$, $\boxed{x^3}$, $\boxed{\sqrt{\square}}$ and $\boxed{\sqrt[3]{\square}}$ keys on a calculator.

$$10 \times 10 = 10^2 = 100 \qquad 10 \times 10 \times 10 = 10^3 = 1000$$
$$\sqrt{100} = 10 \qquad\qquad \sqrt[3]{1000} = 10$$

Squares and square roots

Multiplication	Index notation	Square number	Square root
3×3	3^2	9	$\sqrt{9} = 3$
5×5	5^2	25	$\sqrt{25} = 5$
9×9	9^2	81	$\sqrt{81} = 9$
15×15	15^2	225	$\sqrt{225} = 15$

All numbers, including decimal and negative numbers, can be squared. Only whole-number answers are square numbers (perfect squares).

Only positive numbers have square roots.

$(-4)^2 = -4 \times -4 = 16$

Look at page 10 for negative number rules.

Cubes and cube roots

Multiplication	Index notation	Cube number	Cube root
$2 \times 2 \times 2$	2^3	8	$\sqrt[3]{8} = 2$
$3 \times 3 \times 3$	3^3	27	$\sqrt[3]{27} = 3$
$4 \times 4 \times 4$	4^3	64	$\sqrt[3]{64} = 4$
$5 \times 5 \times 5$	5^3	125	$\sqrt[3]{125} = 5$

All numbers, including decimal and negative numbers, can be cubed. Only whole-number answers are cube numbers (perfect cubes).

All numbers have a cube root.

$\sqrt[3]{216} = 6$

216 is a perfect cube number.

$\sqrt[3]{568} = 8.28$ (2 d.p.)

568 is not a perfect cube.

Worked example

Work out
(a) 8^2 $8 \times 8 = 64$
(b) 12^2 $12 \times 12 = 144$
(c) 1^3 $1 \times 1 \times 1 = 1$
(d) $(-5)^2$ $-5 \times -5 = 25$
(e) $\sqrt{121}$ $121 = 11 \times 11$ so $\sqrt{125} = 11$
(f) $\sqrt[3]{64}$ $64 = 4 \times 4 \times 4$ so $\sqrt[3]{64} = 4$

For part (d) use the cube root function on your calculator. If it is written above the square root button you might need to press the $\boxed{\text{SHIFT}}$ key first.

Worked example

Use a calculator to work out
(a) (i) $17^2 = 289$ (ii) $(-5.4)^2 = 29.16$
(b) (i) $9^3 = 729$ (ii) $3.8^3 = 54.872$
(c) (i) $\sqrt{361} = 19$ (ii) $\sqrt{132.25} = 11.5$
(d) (i) $\sqrt[3]{1331} = 11$ (iii) $\sqrt[3]{-2300}$
                                $= -13.20006...$
                                  $= -13.2$ (1 d.p.)

Write at least 5 decimal places from your calculator display before rounding your answer. For more about rounding decimals see page 3.

Now try this

1 Work out
    (a) $(-2)^3$ (b) $\sqrt{49}$ (c) $\sqrt[3]{1000}$

2 Use a calculator to work out (a) 15^3 (b) $\sqrt{120}$ (c) $\sqrt[3]{-5832}$

3 A farmer has two square fields. The area of field A is $2704\,\text{m}^2$. The area of field B is $2304\,\text{m}^2$. How much longer is field A than field B?

To work out the lengths, find the square roots.

Priority of operations

BIDMAS helps you remember the correct priority of operations. Check if your calculator follows this order automatically. If it does not, you will need to use the brackets keys.

Brackets	$20 - (3 \times 4) = 20 - 12$	brackets then subtraction
Indices (or powers)	$3 \times 4^2 = 3 \times 16 = 48$	index or power first, then multiplication
D/M division and multiplication	$12 + 6 \div 2 = 12 + 3 = 15$	division before addition
A/S addition and subtraction	$16 - 8 \times 2 = 16 - 16 = 0$	subtraction after multiplication

Worked example

Work out

(a) $15 - 5 \times 2$ $\quad = 15 - 10 = 5$
(b) $5 \times 0 - 3$ $\quad = 0 - 3 = -3$
(c) $4 \times 5 \div 10$ $\quad = 20 \div 10 = 2$
(d) $(16 - 7) \times 4$ $\quad = 9 \times 4 = 36$
(e) $(15 - 3) \div (3 \times 2)$ $\quad = 12 \div 6 = 2$
(f) $24 - 3 \times 6 + 10$ $\quad = 24 - 18 + 10 = 16$

If you're not sure which operation to do first, write **BIDMAS**. If there are no brackets or indices, then start with division and multiplication.

Be careful. If a calculation only contains addition and subtraction, you work from left to right.
$24 - 18 + 10 = 6 + 10 = 16$

Worked example

Work out the difference between $(4 + 5)^2$ and $4^2 + 5^2$

$(4 + 5)^2 = 9^2 = 81$
$4^2 + 5^2 = 16 + 25 = 41$
$81 - 41 = 40$

Your calculator might perform the correct order of operations. Try entering the calculations into your calculator in one go.

Worked example

(a) Work these out, and then check with your calculator.

(i) $5 \times (\sqrt{36} + 3^2)$ $\quad = 5 \times (6 + 9)$
$\qquad = 5 \times 15 = 75$
(ii) $8^2 - (6 \times \sqrt{81})$ $\quad = 64 - (6 \times 9)$
$\qquad = 10$
(iii) $(4^2 - \sqrt{100})^2$ $\quad = (16 - 10)^2$
$\qquad = 6^2 = 36$
(iv) $\sqrt[3]{5 \times 5^2}$ $\quad = \sqrt[3]{125} = 5$
(v) $\sqrt{7^2 + 51}$ $\quad = \sqrt{49 + 51}$
$\qquad = \sqrt{100} = 10$

(b) Work out the value of
$$\frac{5^3 + \sqrt[3]{5}}{5^2 - 5}$$
Give your answer to 2 decimal places.

$6.335498... = 6.34$ (2 d.p.)

Now try this

1 Work out
   (a) (i) $24 - 3 \times 5$
   $\quad$ (ii) $8 \times 4 \div 2$
   (b) (i) $11^2 - (\sqrt{49} \times \sqrt[3]{125})$
   $\quad$ (ii) $(5^2 - 4^2)^2$
   (c) (i) $\sqrt{11} \times \sqrt{11}$
   $\quad$ (ii) $\sqrt{19} \times \sqrt{19}$

2 Mike says that $(5 \times 6)^2$ gives the same answer as $5^2 \times 6^2$. Show that he is correct.

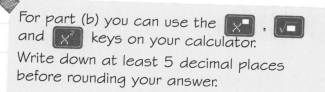

For part (b) you can use the  and x^2 keys on your calculator.
Write down at least 5 decimal places before rounding your answer.

More powers

In numbers such as 2^2, 2^3 and 10^4 the small raised number is the **power** or **index**.

Turn to page 6 for a reminder about powers of 10 and to page 12 for a reminder about squares and cubes.

Powers can be higher than 2 and 3. Powers can also be negative.
The power tells you how many of the number to multiply together. $7^4 = 7 \times 7 \times 7 \times 7$

The index keys on your calculator may look like , , , or .
Make sure you know how to use them.

Worked example

Write each product as a single power.
(a) $6^2 \times 6^7 = 6^{2+7} = 6^9$
(b) $9^5 \div 9^2 = 9^{5-2} = 9^3$
(c) $5^6 \div 5 = 5^{6-1} = 5^5$
(d) $(10^3)^2 = 10^3 \times 10^3 = 10^{3+3} = 10^6$

Calculator examples

7^4:   7 4 = 2401

5.6^5:   5.6 ▣ 5 = 5507.31776

$(-10)^3$:   (−) 10 ▣ 3 = −1000

10^{-3}:   10 ▣ (−) 3 = $\dfrac{1}{1000}$

$\left(\dfrac{1}{4}\right)^3$:   1 ÷ 4 ▣ 3 = $\dfrac{1}{64}$

5^0:   5 ▣ 0 = 1

Golden rules

- To multiply powers of the same number, add the powers.
  $5^4 \times 5^2 = 5^{4+2} = 5^6$

- To divide powers of the same number, subtract the powers.
  $5^4 \div 5^2 = 5^{4-2} = 5^2$

- To work out powers of terms in brackets, multiply the powers.
  $(5^4)^3 = 5^{3 \times 4} = 5^{12}$

- If there is no index, the power is 1.

- Any number to the power 0 is 1.

Worked example

Use the index key on your calculator to work these out.
(a) $5^6 = 15\,625$ (b) $4.5^4 = 410.0625$
(c) $(-5)^3 = -125$ (d) $5^{-3} = 0.008$
(e) $\left(\dfrac{1}{3}\right)^3 = 0.037$ (f) $8^0 = 1$

Worked example

(a) Write 6^{-2} as a fraction. $\dfrac{1}{6^2} = \dfrac{1}{36}$
(b) Work out:
  (i) $\dfrac{1}{8^2} = \dfrac{1}{8 \times 8} = \dfrac{1}{64}$

  (ii) $\dfrac{1}{10^4} = \dfrac{1}{10 \times 10 \times 10 \times 10}$
          $= \dfrac{1}{1000}$

  (iii) $2^{-4} = \dfrac{1}{2^4} = \dfrac{1}{2 \times 2 \times 2 \times 2} = \dfrac{1}{16}$

  (iv) $10^{-4} = \dfrac{1}{10^4} = \dfrac{1}{10000}$

Negative powers

$a^{-n} = \dfrac{1}{a^n}$ $5^{-2} = \dfrac{1}{5^2} = \dfrac{1}{25}$

Be careful!

A **negative** power can have a **positive** answer.

Now try this

1 Write each product as a single power.
  (a) $4 \times 4 \times 4 \times 4 \times 4 \times 4$ (b) $8^3 \times 8^5$
  (c) $10^7 \div 10^4$ (d) $10 \times 10^6 \div 10^2$

2 Work out the value of
  (a) $\dfrac{1}{5^2}$ (b) 3^{-2} (c) 12^0

3 Use a calculator to work out
  (a) 5^4 (b) $(-6)^3$ (c) 5^{-3}

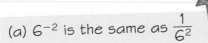

(a) 6^{-2} is the same as $\dfrac{1}{6^2}$

Prime factors

All numbers can be written as a product of prime factors. This is sometimes called **prime factor decomposition**.

For a reminder about factors, multiples and prime numbers, have a look at page 11.

Prime numbers

A prime number is a number that only has **two** factors, 1 and itself.

All numbers bigger than 1 that are **non-prime** have three or more factors. They are called **composite** numbers.

Prime factors

You can write any composite number as a product of its prime factors.

The factors of 12 are 1, 2, 3, 4, 6 and 12.

The prime factors of 12 are 2 and 3. You can also use powers to write the product.

$12 = 2 \times 2 \times 3 = 2^2 \times 3$

Worked example

(a) Match each number to the product of its prime factors.

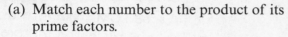

36   66   70   100

$2 \times 5 \times 7$ $2 \times 3 \times 11$ $2^2 \times 5^2$ $2^2 \times 3^2$

(b) Work out $2 \times 3^2 \times 5^3$

= 2250

(a) Write $2^2 \times 5^2$ as $2 \times 2 \times 5 \times 5$

(b) Use the index keys on your calculator, or work out $2 \times 9 \times 125$.

Worked example

Write 120 as a product of its prime factors.

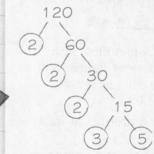

$120 = 2 \times 2 \times 2 \times 3 \times 5 = 2^3 \times 3 \times 5$

Follow these steps:

1. Write down the prime numbers between 1 and 20.

2. Draw a factor tree. Think about the numbers that multiply to make 120.

3. Circle the prime factors.

4. Remember to check that all the factors are prime numbers.

You can start to work out the prime factor decomposition with any factor pair of the number.

Write the number as a product of all four factors.

Write the factors in numerical order.

Use index notation for $5 \times 5 = 5^2$

Check it!

$2 \times 3 \times 5^2 = 2 \times 3 \times 25$

$= 6 \times 25 = 150$ ✓

Worked example

Complete the prime factor tree for 150 and then write 150 as a product of its prime factors.

$10 = 2 \times 5; \ 15 = 3 \times 5$ 150

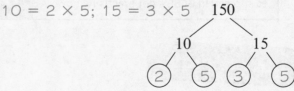

$150 = 2 \times 5 \times 3 \times 5$
$150 = 2 \times 3 \times 5^2$

Now try this

1 Draw and complete a prime factor tree to write 180 as a product of prime factors.

2 Write each number as a product of prime factors. Rewrite, using index notation.
   (a) 24   (b) 64   (c) 140   (d) 200

Start with 10 × 18

HCF and LCM

You can use prime factor decomposition to find the HCF and LCM of sets of numbers.

For a reminder about HCF and LCM turn to page 11, and for prime factor decomposition turn to page 15.

Worked example

Complete the factor trees and write the prime factor decompositions for 36 and 54. Use these to work out the HCF and LCM of 36 and 54.

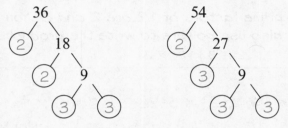

$36 = \underline{2} \times 2 \times \underline{3} \times \underline{3}$
$54 = \underline{2} \times \underline{3} \times \underline{3} \times 3$
$HCF = 2 \times 3 \times 3 = 18$
$LCM = 18 \times 2 \times 3 = 108$

1. Complete the prime factor trees.

2. Write 36 and 54 as products **without index notation**.

3. Underline the common factors.

4. To find the **HCF** of 36 and 54, multiply the common factors together.

5. To find the **LCM**, multiply the HCF by the remaining factors.

For an alternative method of finding the HCF and LCM, turn to page 11.

Problem solved!

The total number of brownies must be a multiple of 30. The number of flapjacks must be a multiple of 45. You can answer this question by finding the LCM.

- Write each number as a product of its prime factors, without indices.
- Multiply together the factors common to both products to find the HCF.
- Multiply the HCF by the remaining factors to find the LCM.
- Work out how many batches of each type of product the bakery needs to make.

> You'll need brilliant problem-solving skills to succeed in GCSE – get practising now!

Worked example

A bakery makes brownies in batches of 30 and flapjacks in batches of 45.

What is the lowest number of batches of each type of product the bakery should make to have the same number of each type?

$30 = 2 \times \underline{3} \times \underline{5}$
$45 = 3 \times \underline{3} \times \underline{5}$
$HCF = 3 \times 5 = 15$
$LCM = 15 \times 2 \times 3 = 90$
They need 90 of each type.
$90 \div 30 = 3, 90 \div 45 = 2$
3 batches of brownies and 2 batches of flapjacks

Now try this

1 $32 = 2^5$ and $36 = 2^2 \times 3^2$. Work out the HCF and LCM of 32 and 36.

2 Find the HCF and LCM of 72 and 120.

> Work out the HCF of 32 and 48.

3 Sue has two rolls of coloured tape. One is 48 m long and the other is 32 m. She wants to cut both of them into pieces of the same length so that no tape is left over. What is the longest length she can cut them into?

> Work out the LCM of 24 and 18.

4 A baker makes large and small bread rolls. He can fit 24 small rolls and 18 large rolls on his baking trays. He wants to make the same number of each size. What is the minimum number of trays he can make of each size?

Standard form

Standard form is a useful way of writing very large or very small numbers.

5 674 000 = 5.674 million   a number between 1 and 10   a power of 10

$$5\,674\,000 = 5.674 \times 10^6$$

The power may be positive or negative.

$$73\,000 = 7.3 \times 10^4$$

numbers ⩾ 1 have a positive power

$$0.00876 = 8.76 \times 10^{-3}$$

numbers < 1 have a negative power

For a reminder about powers of 10 and millions as a decimal turn to page 6.

Worked example

Write each product as an ordinary number.
(a) 4.5×10^3 = 4500
(b) 6.78×10^2 = 678
(c) 8.9×10^{-3} = 0.0089
(d) 5.25×10^{-2} = 0.0525

Golden rule

Multiplying by a negative power is the same as dividing by a positive power.

$3.5 \times 10^{-2} = 3.5 \div 10^2$
$\phantom{3.5 \times 10^{-2}} = 3.5 \div 100 = 0.035$

8.9×10^{-3} $= 8.9 \div 10^3$
$\phantom{8.9 \times 10^{-3}}$ $= 8.9 \div 1000 = 0.0089$

Numbers in standard form

Count **decimal places** to convert ordinary numbers to standard form.

Imagine that the decimal point jumps along the number.

The power tells you how many places to the right (positive) or to the left (negative) the decimal point jumps.

4 jumps left, so power is +4

$73\,000 \quad = \quad 7.3 \times 10^4$

3 jumps right, so power is −3

$0.00876 \quad = \quad 8.76 \times 10^{-3}$

$73\,000 \geqslant 1$ so the power is positive.
$0.00876 < 1$ so the power is negative.

Worked example

(a) Write these numbers in standard form.
  (i)   4800 $= 4.8 \times 1000 = 4.8 \times 10^3$
  (ii) 0.456 $= 4.56 \div 10 = 4.56 \times 10^{-1}$
  (iii) 0.0072 $= 7.2 \div 1000 = 7.2 \times 10^{-3}$
(b) Write these as ordinary numbers.
  (i) 6.3×10^4 $= 63\,000$
  (ii) 8.07×10^2 $= 807$
  (iii) 3.246×10^{-2} $= 0.03246$
  (iv) 8.9×10^{-1} $= 0.89$

Look at the power to see how many places the decimal point jumps.

The first part of each standard form number must be a number greater than or equal to 1 and less than 10. Work out the power of 10 which you would need to multiply this number by to get the original number.

Now try this

1 Write each of these as an ordinary number:
  (a) 7.5×10^3 (b) 6.4×10^{-2}

2 Write these numbers in standard form:
  (a) 5600   (b) 0.099

3 Which number is larger, 3.4×10^5 or 9.9×10^4?

Calculator buttons

A **scientific calculator** can work out very complicated calculations, including fractions. Knowing how to use some of these calculator buttons will help you to save time.

fraction buttons   index buttons   convert between fractions and decimals

For a reminder about index buttons turn to pages 12–14.

Worked example

Match each button to its function.

Add
Delete last digit or operation
Square the number entered
Clear the display
Take square root of number entered
Equals or enter for the answer
Take cube root of number entered
Change fraction to decimal

For a reminder about the x^y, $\sqrt{\square}$ and $\sqrt[3]{\square}$ buttons turn to page 12.

$+$, $-$, $\times$ and $\div$ are the operations buttons for add, subtract, multiply and divide.

$S\!\Leftrightarrow\!D$ is a useful key if you want an answer in decimal form; it is often on the right-hand side above the DEL and AC keys.

$=$ is used for completing the calculation.

Fraction buttons

Use these three buttons to enter fractions and mixed numbers into a calculator.

Up
Left ◀ REPLAY ▶ Right   Use this button to move out of current function.
Down

 → Fraction key
Enter numerator, down arrow, enter denominator, right arrow

To enter $\frac{2}{3}$:

 → Mixed number key
Enter whole number, right arrow, enter numerator, down arrow, enter denominator, right arrow

To enter $2\frac{3}{4}$:

  2 3 4

Worked example

Use your calculator to:
(a) change $\frac{4}{5}$ into a decimal   0.8
(b) change $3\frac{6}{7}$ into an improper fraction $\frac{27}{7}$
(c) change $\frac{23}{6}$ into a mixed number $3\frac{5}{6}$
(d) work out $\frac{1}{4} + \frac{2}{5}$ $\frac{13}{20}$

(a) Use the button to enter $\frac{4}{5}$, then $=$, then $S\!\Leftrightarrow\!D$

(b) Use the ▦ button to enter $3\frac{6}{7}$, then $=$

(c) Use the ▦ button to enter $\frac{23}{6}$, then $=$, and then SHIFT 2nd function $S\!\Leftrightarrow\!D$

(d) Use the ▦ button to enter $\frac{1}{4}$, then  $+$ ▦ to enter $\frac{2}{5}$, ◀ $=$

Now try this

1 Change
  (a) $\frac{7}{20}$ into a decimal
  (b) $2\frac{3}{4}$ into an improper fraction
  (c) $\frac{42}{8}$ into a mixed number.

2 Harry forgets to use the right arrow button after entering $\frac{4}{5}$ when working out $\frac{4}{5} - \frac{2}{3}$.
  (a) Try what Harry did on your calculator. Write down the calculation shown on the display.
  (b) Now calculate $\frac{4}{5} - \frac{2}{3}$ correctly.

Fraction basics

To compare and calculate with fractions successfully, you must be able to recognise, use and simplify **equivalent fractions**. You also need to be able to find **fractions of amounts**.

① Parts of a fraction

You can use fractions to divide an object into parts.

The top number is called the **numerator**. → $\frac{3}{4}$ of this rectangle is shaded.

The bottom number is called the **denominator**.

② Equivalent fractions

Different fractions can describe the same amount.

$\frac{2}{3}$
$\frac{6}{9}$

$\overset{\times 3}{\frac{2}{3} = \frac{6}{9}}$
$\underset{\times 3}{}$

You can find equivalent fractions by multiplying or dividing the numerator and denominator by the same number.

③ Simplifying fractions

To **cancel** or **reduce** a fraction you divide the top and bottom by the same number.

$\overset{\div 20}{\frac{40}{60}} = \frac{2}{3}$
$\underset{\div 20}{}$

Divide by the HCF of the numerator and denominator to get the fraction in its **simplest form**.

Look back at page 11 for a reminder about HCF.

④ Fractions of amounts

You can find fractions of amounts.

| £60 | £60 | £60 | £60 |

←——— £240 ———→

To work out $\frac{3}{4}$ of £240:

£240 ÷ 4 = £60 —divide by the denominator

60 × 3 = £180 —multiply by the numerator

$\frac{3}{4}$ of £240 = £180

(a) Divide both parts by 7 (the HCF of 35 and 42).

(b) and (c) To compare or order fractions, convert them to equivalent fractions with the same denominator (the LCM of the denominators of the fractions you are comparing).

In (c) remember to write the original fractions in order.

Worked example

(a) Write $\frac{35}{42}$ in its simplest form. $\frac{5}{6}$

(b) Write < or > between the fractions in each pair, to compare them.

(i) $\frac{4}{5}$ and $\frac{7}{10}$ $\frac{4}{5} = \frac{8}{10}$ so $\frac{4}{5} > \frac{7}{10}$

(ii) $\frac{3}{8}$ and $\frac{5}{12}$ $\frac{3}{8} = \frac{9}{24}$ and $\frac{5}{12} = \frac{10}{24}$ so $\frac{3}{8} < \frac{5}{12}$

(c) Order these fractions, smallest to largest.

$\frac{1}{2}$ $\frac{5}{8}$ $\frac{2}{5}$ $\frac{3}{4}$

$\frac{1}{2} = \frac{20}{40}, \frac{5}{8} = \frac{25}{40}, \frac{2}{5} = \frac{16}{40}, \frac{3}{4} = \frac{30}{40}$

$\frac{2}{5}, \frac{1}{2}, \frac{5}{8}, \frac{3}{4}$

Worked example

Work out $\frac{3}{5}$ of £150

150 ÷ 5 = 30

30 × 3 = 90

$\frac{3}{5}$ of £150 is £90   150 ÷ 5 × 3

Work out $\frac{1}{2}$ of £250 and $\frac{2}{5}$ of £250. Subtract these amounts from £250 to find how much third place receives.

Now try this

1 Simplify

(a) $\frac{30}{90}$ (b) $\frac{12}{48}$

2 Use equivalence to order these fractions, smallest to largest: $\frac{4}{5}$ $\frac{1}{2}$ $\frac{7}{10}$ $\frac{15}{20}$

3 Prize money of £250 is shared in this way: $\frac{1}{2}$ to the winner, $\frac{2}{5}$ to second place and the rest to third place. How much does each one get?

Changing fractions

Before you can calculate with fractions, you need to know how to change between mixed numbers and improper fractions.

A **mixed number** has a whole number part and a fraction part.

$$1 \quad + \quad \tfrac{3}{4} \quad = \quad 1\tfrac{3}{4}$$

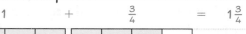

whole-number part ← $1\tfrac{3}{4}$ → fraction part

In an **improper fraction** the numerator is larger than the denominator.

$$\tfrac{4}{4} \quad + \quad \tfrac{3}{4} \quad = \quad \tfrac{7}{4}$$

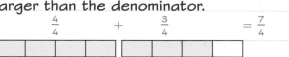

$\tfrac{7}{4}$ ← numerator > denominator

Golden rules

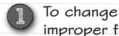

1 To change a mixed number into an improper fraction:

- multiply the whole-number part by the denominator of the fraction
- add the numerator of the fraction part.

$$2\tfrac{3}{4} = \frac{2 \times 4 + 3}{4} = \frac{11}{4}$$

The denominator is the same as the fraction denominator.

2 To change an improper fraction into a mixed number:

- divide the numerator by the denominator
- write the quotient as the whole number
- write the remainder as the numerator
- keep the denominator the same.

fraction   whole number   remainder

$$\frac{31}{6} = 31 \div 6 = 5 \text{ r } 1 = 5\tfrac{1}{6}$$

Worked example

(a) Change $4\tfrac{5}{6}$ into an improper fraction.

$$4\tfrac{5}{6} = \frac{4 \times 6 + 5}{6} = \frac{29}{6}$$

(b) Change $\tfrac{33}{9}$ into a mixed number:

$$\frac{33}{9} = 33 \div 9 = 3 \text{ r } 6 = 3\tfrac{6}{9} = 3\tfrac{2}{3}$$

(a) Multiply the whole number by the denominator and then add the numerator.

(b) Divide the numerator by the denominator. Simplify the fraction.

For a reminder about simplifying look at page 19.

Problem solved!

(a) Change $8\tfrac{1}{3}$ into an improper fraction or $\tfrac{26}{3}$ into a mixed number.

(b) Change $5\tfrac{1}{2}$ to an improper fraction and then change to an equivalent fraction with a denominator of 4.

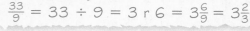

You'll need brilliant problem-solving skills to succeed in GCSE – get practising now!

Worked example

(a) Which is larger, $8\tfrac{1}{3}$ or $\tfrac{26}{3}$?

$8\tfrac{1}{3} = \tfrac{25}{3}$ and $\tfrac{26}{3} = 8\tfrac{2}{3}$

So $\tfrac{26}{3}$ is larger.

(b) How many $\tfrac{1}{4}$s are there in 5?

$$5\tfrac{1}{2} = \frac{10 + 1}{2} = \frac{11}{2} = \frac{22}{4}$$

There are 22 $\tfrac{1}{4}$s in $5\tfrac{1}{2}$

Now try this

1 Change $5\tfrac{5}{9}$ into an improper fraction.

2 Change $\tfrac{45}{6}$ into a simplified mixed number.

3 Which is smaller, $4\tfrac{3}{5}$ or $\tfrac{21}{5}$?

Change $5\tfrac{1}{2}$ into an improper fraction in **sixths**. $\tfrac{1}{2} = \tfrac{3}{6}$

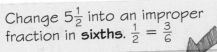

4 Hollie has to cut $5\tfrac{1}{2}$ cakes into sixths.
How many pieces will she have?

Add and subtract fractions

If you need to add or subtract fractions with different denominators, write them as equivalent fractions with the same denominator.
Here are two examples:

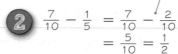
Add the numerators

Subtract the numerators

 ① $\frac{1}{3} + \frac{3}{4} + \frac{5}{6} = \frac{4}{12} + \frac{9}{12} + \frac{10}{12}$
$= \frac{23}{12} = 1\frac{11}{12}$

② $\frac{7}{10} - \frac{1}{5} = \frac{7}{10} - \frac{2}{10}$
$= \frac{5}{10} = \frac{1}{2}$

Find equivalent fractions with the same denominator. You can use 12 as the denominator (the LCM of 3, 4 and 6).

10 is a multiple of 5, so you can use 10 as the denominator.

 Turn to page 19 for a reminder about equivalent fractions.

Worked example

Work out

(a) $\frac{2}{3} + \frac{4}{5}$ $= \frac{10}{15} + \frac{12}{15} = \frac{22}{15} = 1\frac{7}{15}$

(b) $\frac{8}{9} - \frac{5}{6}$ $= \frac{16}{18} - \frac{15}{18} = \frac{1}{18}$

(c) $\frac{3}{4} + \frac{5}{12} - \frac{2}{3}$ $= \frac{9}{12} + \frac{5}{12} - \frac{8}{12} = \frac{6}{12} = \frac{1}{2}$

(a) Convert the final answer to a mixed number.

(b) The LCM of 9 and 6 is 18.

(c) Simplify the final answer if possible.

Problem solved!

If you can't see what calculation to do, try simpler numbers. If Lucie had worked for 15 minutes, and Aimee had worked for 10 minutes, how much longer would Lucie have worked? The answer is $15 - 10$. So you need to work out $\frac{5}{6} - \frac{5}{8}$.

You'll need brilliant problem-solving skills to succeed in GCSE – get practising now!

Worked example

Lucie and Aimee have the same homework task. Lucie has completed $\frac{5}{6}$ of the task and Aimee has completed $\frac{5}{8}$. How much more has Lucie completed than Aimee?

$\frac{5}{6} - \frac{5}{8} = \frac{20}{24} - \frac{15}{24} = \frac{5}{24}$

Worked example

Work out $3\frac{2}{5} - 1\frac{3}{4}$

$3 - 1 = 2$ and $\frac{8}{20} - \frac{15}{20} = -\frac{7}{20}$
$2 - \frac{7}{20} = 1\frac{13}{20}$

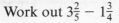 Work out the whole-number subtraction and work out the fraction subtraction, changing denominators to 20. Combine the two parts.

Now try this

1 Work out $\frac{3}{4} + \frac{11}{12}$. Give your answer as a simplified mixed number.

2 Work out $4\frac{5}{6} + 3\frac{2}{5}$

3 Ben has one blue cable and one red cable. The blue cable is $3\frac{2}{5}$ m long and the red cable is $2\frac{2}{3}$ m long.
How much longer is the blue cable?

 Work out $3\frac{2}{5} - 2\frac{2}{3}$, following the steps above. Be careful, $\frac{2}{5} < \frac{2}{3}$, so the subtraction of the fraction parts will give a negative answer.

Multiply and divide fractions

Learn the rules for multiplying and dividing fractions – the method is different from adding and subtracting. Remember to give answers in their simplest form.

Look back at page 19 for a reminder on simplifying fractions.

Multiplying fractions

Follow these steps to multiply fractions:

1 Multiply the numerators together and then multiply the denominators together.

$$\frac{3}{5} \times \frac{4}{9} = \frac{3 \times 4}{5 \times 9} = \frac{12}{45}$$

2 Write any whole numbers as fractions with a denominator of 1.

$$5 \times \frac{3}{8} = \frac{5}{1} \times \frac{3}{8} = \frac{5 \times 3}{1 \times 8} = \frac{15}{8} = 1\frac{7}{8}$$

3 You can sometimes cancel the fractions before multiplying.

$$\frac{2}{3} \times \frac{9}{10} = \frac{^1 2}{_1 3} \times \frac{9^3}{10_5} = \frac{1}{1} \times \frac{3}{5} = \frac{3}{5}$$

Convert $1\frac{2}{3}$ and $3\frac{1}{5}$ into improper fractions, then multiply the numerators and multiply the denominators. Cancel if possible.

For a reminder about working out fractions on a calculator, see page 18.

Golden rule

When multiplying or dividing with **mixed numbers** convert them to improper fractions before doing any calculations.

Worked example

Work these out. Check your answers with the fraction buttons on your calculator.

(a) $\frac{8}{9} \times \frac{5}{6} = \frac{8 \times 5}{9 \times 6} = \frac{40}{54} = \frac{20}{27}$

(b) $1\frac{2}{3} \times 3\frac{1}{5} = \frac{\cancel{5} \times 16}{3 \times \cancel{5}_1} = \frac{1 \times 16}{3 \times 1} = \frac{16}{3}$
$= 5\frac{1}{3}$

Dividing fractions

Follow these steps to divide fractions:

1 Write any whole numbers as fractions with a denominator of 1.

2 Change ÷ to × and turn the second fraction upside down (use the reciprocal).

3 Multiply as normal.

$$6 \div \frac{2}{3} = \frac{6}{1} \times \frac{3}{2} = \frac{^3 6 \times 3}{1 \times 2_1} = 9$$

$$\frac{4}{5} \div \frac{1}{4} = \frac{4}{5} \times \frac{4}{1} = \frac{16}{5} = 3\frac{1}{5}$$

Reciprocals

When you find a reciprocal, the numerator becomes the denominator and the denominator becomes the numerator.

$$\frac{3}{4} \rightarrow \frac{4}{3}$$

Write a whole number as a fraction with a denominator of 1 before turning it upside down. If a fraction has a numerator of 1, its reciprocal will be a whole number.

$$6 \rightarrow \frac{6}{1} \qquad \frac{1}{5} \rightarrow \frac{5}{1}$$

Worked example

Work these out. Check your answers with the fraction buttons on your calculator.

(a) $\frac{8}{9} \div \frac{5}{6} = \frac{8}{9} \times \frac{6}{5} = \frac{8 \times 6}{9 \times 5} = \frac{^{16}\cancel{48}}{\cancel{45}_{15}} = \frac{16}{15} = 1\frac{1}{15}$

(b) $3\frac{3}{5} \div 2\frac{1}{10} = \frac{^6\cancel{18}}{_1\cancel{5}} \times \frac{\cancel{10}^2}{\cancel{21}_7} = \frac{6 \times 2}{1 \times 7} = \frac{12}{7} = 1\frac{5}{7}$

(a) Use the reciprocal of $\frac{5}{6}$ and change ÷ to ×.
(b) Convert mixed numbers to improper fractions first.

Now try this

1 Work out $\frac{3}{4} \times \frac{11}{12}$. Give your answer in its simplest form.

2 Work out $4\frac{1}{6} \times 3\frac{3}{5}$

3 Lynne has a $4\frac{4}{5}$m length of rope to cut into six equal pieces.
What is the length of each piece?

$$4\frac{4}{5} \div 6 = 4\frac{4}{5} \div \frac{6}{1}$$

Fractions, division, decimals

You need to know the link between fractions and division, how to change a fraction into a decimal and the difference between **terminating** and **non-terminating** decimals.

Look at pages 8, 9 and 19 for reminders about division and finding fractions of amounts.

terminating decimal – this has 3 decimal places	recurring decimal (a digit or group of digits repeats forever)	non-terminating (carries on forever)
5.125	4.555... = 4.$\dot{5}$	3.141 59...

Worked example

Use division to work out

(a) $\frac{1}{4}$ of 2349 $4\overline{)2^23^34^29.^10^20}$ = 587.25

(b) $\frac{1}{8}$ of 2349 $8\overline{)2^23^74^29.^50^20^40}$ = 293.625

(c) $\frac{1}{9}$ of 3125 $9\overline{)3^31^42^65.^20^20^20}$ = 347.$\dot{2}$

(d) $\frac{1}{6}$ of 2315 $6\overline{)2^23^51^35.^50^20^20}$ = 385.8$\dot{3}$

> To find $\frac{1}{4}$ of 2349 work out 2349 ÷ 4. You might need to add extra zeros after the decimal point.

> If the answer is a recurring decimal, the remainders will start to repeat. In part (c) the 2s in the answer will go on forever.

To convert a fraction to a decimal, divide the numerator by the denominator. You might have to press the S⇒D button.

When a **group of digits** recur you put dots over the first and last digits:

$\frac{4}{7}$ = 0.$\dot{5}$71 42$\dot{8}$

= 0.571 428 571 428...

Worked example

Use your calculator and match each fraction with its decimal equivalent.

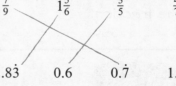

$\frac{7}{9}$ $1\frac{5}{6}$ $\frac{3}{5}$ $\frac{4}{7}$ $1\frac{4}{5}$

1.8$\dot{3}$ 0.6   0.$\dot{7}$ 1.8   0.$\dot{5}$71 42$\dot{8}$

Use dot notation for recurring decimals.

Now try this

1 Use your calculator to change these fractions to decimals.
   (a) $\frac{3}{7}$ (b) $\frac{5}{12}$

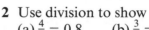
2 Use division to show
   (a) $\frac{4}{5}$ = 0.8   (b) $\frac{3}{8}$ = 0.375

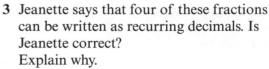
3 Jeanette says that four of these fractions can be written as recurring decimals. Is Jeanette correct?
   Explain why.

   $\frac{3}{4}$ $\frac{2}{9}$ $\frac{1}{3}$ $\frac{1}{6}$ $\frac{5}{8}$ $\frac{7}{12}$

Worked example

Use division to change each fraction into a decimal.

(a) $\frac{1}{8}$ $8\overline{)1.^10^20^40}$ = 0.125

(b) $\frac{2}{3}$ $3\overline{)2.^20^20^20}$ = 0.$\dot{6}$

(c) $\frac{5}{9}$ $9\overline{)5.^50^50^50}$ = 0.$\dot{5}$

Equivalence

You need to be able to change (convert) between fractions, decimals and percentages. Learn these useful equivalents.

Fraction	$\frac{1}{10}$	$\frac{3}{10}$	$\frac{1}{100}$	$\frac{1}{5}$	$\frac{2}{5}$	$\frac{1}{4}$	$\frac{1}{2}$	$\frac{3}{4}$	$\frac{1}{3}$
Decimal	0.1	0.3	0.01	0.2	0.4	0.25	0.5	0.75	$0.333 \ldots = 0.\dot{3}$
Percentage	10%	30%	1%	20%	40%	25%	50%	75%	$33.333\ldots = 33.\dot{3}\%$

Conversions

Learn these methods for converting between fractions, decimals and percentages:

1 DECIMAL ⇄ PERCENTAGE (%)
   × 100 → ÷ 100

$0.75 \times 100 = 75\%$ $7\% \div 100 = 0.07$

2 Write a **decimal** as a fraction with a denominator of 10 or 100, and then simplify the fraction.

$0.9 = \frac{9}{10}$ $0.12 = \frac{12}{100} = \frac{3}{25}$

3 Write a **percentage** as a fraction with a denominator of 100, and then simplify the fraction.

$80\% = \frac{80}{100} = \frac{4}{5}$ $35\% = \frac{35}{100} = \frac{7}{20}$

Worked example

Complete the table showing equivalent fractions, decimals and percentages.

Fraction	Decimal	Percentage
$\frac{7}{10}$	0.7	70%
$\frac{22}{25}$	0.88	88%
$\frac{1}{20}$	0.05	5%
$\frac{4}{5}$	0.8	80%
$\frac{11}{100}$	0.11	11%
$\frac{13}{20}$	0.65	65%

Make sure you simplify fractions:
$88\% = \frac{88}{100} = \frac{22}{25}$

You could also use equivalent fractions:
$\frac{17}{20} = \frac{85}{100}$ (multiply both numbers by 5)

Worked example

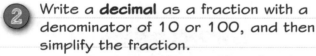

Bob scores 17 out of 20 in a test. What is his score as a percentage?

$17 \div 20 = 0.85 = 85\%$

Convert fractions and percentages to decimals.
$56\% = 0.56$ $\frac{11}{20} = \frac{55}{100} = 0.55$

Worked example

Write these in order of size. Start with the smallest value.

   0.5   56% $\frac{11}{20}$ 0.49

0.49   0.5 $\frac{11}{20}$ 56%

Now try this

1 As a percentage, which score is higher, 21 out of 25 or 17 out of 20?

2 Write these in order of size. Start with the smallest value.

   0.26   6 out of 20 $\frac{1}{4}$ $\frac{1}{5}$ 24%

Convert so that all values are decimals or percentages, order these, and write them out again using the original form in the question.
6 out of 20 $= \frac{6}{20}$

Percentages

Per cent means 'out of 100'. Percentages play an important role in real life, so it is vital that you can find percentages of amounts.

$\frac{1}{10} = 10\%$

100%										1 whole = 100%
10%	10%	10%	10%	10%	10%	10%	10%	10%	10%	10 × 10% = 100%
20%		80%								20% + 80% = 100%

20% = 2 × 10%   80% = 8 × 10%

Percentages of amounts

 1 To find a percentage of an amount:
- divide the amount by 100 to find 1%
- multiply this 1% by the percentage.

  65% of £1250 = 1250 ÷ 100 × 65
   = £812.50

  12.5% of 1640 km = 1640 ÷ 100 × 12.5
   = 205 km

2 To write one quantity as a percentage of another:
- divide the first quantity by the second
- multiply the result by 100.

  26 out of 40: 26 ÷ 40 × 100 = 65%
  4 out of 32: 4 ÷ 32 × 100 = 12.5%

Non-calculator methods

Use these rules to find percentages without using a calculator.

To find	Work out	To find	Work out
10%	÷ 10	50%	÷ 2
1%	÷ 100	25%	÷ 4
5%	10% ÷ 2	75%	50% + 25%
20%	10% × 2		

You can use combinations of simpler percentages to find harder percentages. For example, to find 64% of £360:

64% = 50% + 10% + (4 × 1%)
 = 180 + 36 + 4 × 3.6
 = £230.40

Worked example

(a) Work out 38% of £1580.

1580 ÷ 100 × 38 = 600.4
£600.40

(b) Express 24 out of 60 as a percentage.

24 ÷ 60 × 100 = 40
40%

(c) Work out 15% of £240.

10% = 24; 5% = 12
15% = 10% + 5% = 24 + 12 = £36

Worked example

(a) If 20% of a number is 16, what is the number?

20% is 16 so 10% is 16 ÷ 2 = 8
100% = 10 × 8 = 80
The number is 80

(b) Of 120 members of a gym club, 72 are under 12 years old. What percentage of the club members is 12 years old or older?

72 ÷ 120 × 100 = 60% (under 12 years)
100% − 60% = 40%
40% are aged 12 or older

Now try this

In a competition, the prize money of £640 is shared between the first **four** winners. The first winner gets 50%, the second winner gets 30%, and the third winner gets 15%. How much does each of the **four** winners get?

Work out the value of the first three prizes and subtract from £640 to find the value of the fourth.

Number problem-solving

To answer problem-solving and multi-step questions you will need to apply your mathematical skills and your knowledge. Read the question carefully and write down all the calculations you need. You may pick up method marks, even if you do not finish the question.

For reminders on addition, subtraction and multiplication turn to pages 4, 5 and 7. For fractions and percentages turn to pages 19 and 25.

Worked example

Ami buys 4 coffees, 2 teas, 3 squashes, 4 soups, 5 sandwiches and 2 bread rolls. She pays with two £20 notes and a £10 note. How much change should she get?

Joe's cafe	
Coffee	£2.25
Tea	£1.75
Squash	99p
Soup	£3.60
Bread roll	80p
Sandwich	£3.05

Coffee: 2.25 × 4 = £9
Tea: 2 × £1.75 = £3.50
Squash: 3 × 99p = £2.97
Soup: 4 × £3.60 = £14.40
Sandwiches: 5 × £3.05 = £15.25
Bread rolls: 2 × 80p = £1.60
£9 + £3.50 + £2.97 + £14.40 + £15.25
    + £1.60 = £46.72
Ami pays £20 + £20 + £10 = £50
£50 − £46.72 = £3.28

Ami should get £3.28 change.

Problem solved!

Write words with your working to show what you are calculating at each stage.
Work out the cost of 4 coffees, 2 teas, and so on.
Then add up the totals to work out Ami's total bill.
Work out how much Ami gave the cashier.
Finally subtract to find her change.

You'll need brilliant problem-solving skills to succeed in GCSE – get practising now!

Worked example

Of the 240 students a school has in Y9, $\frac{3}{8}$ study French, 25% study German, $\frac{3}{10}$ study Spanish and the rest study Russian. Each student studies only one language.
(a) How many students study Russian?

$\frac{3}{8}$ of 240 = 90

25% of 240 = 60

$\frac{3}{10}$ of 240 = 72

90 + 60 + 72 = 222

240 − 222 = 18

(b) What fraction of students study Russian? Give your answer in its simplest form.

$\frac{18}{240} = \frac{3}{40}$

Problem solved!

Write all your calculations.
(a) Work out all the fractions and percentages of 240. Add these together. Subtract from 240.
(b) Write the answer to part (a) as a fraction out of 240 and then simplify the fraction.

You'll need brilliant problem-solving skills to succeed in GCSE – get practising now!

Now try this

1 Here is part of Jackie's electricity bill. How much does she have to pay, to the nearest penny?

Old reading	32789 units
New reading	33018 units
Cost per unit	9.35 pence

2 Three friends share a bill of £43. Sonia says that they each need to pay £14.30. Show that she is incorrect.

Subtract the old reading from the new reading to find how many units she has used. Multiply this by the cost per unit. Change your answer to pounds and pence. Round the pence to the nearest whole penny.

Collecting like terms

In algebra you use letters to represent unknown numbers.

An **expression** contains letters and numbers, but no equals sign.

this is an expression

$$5x - 2y + 7$$

each of these is a term

You can simplify expressions by **collecting like terms.**

$c + c + c + c = 4c$

    this means 4 lots of c or $4 \times c$

$$3x + 2y - x + 5y = 3x - x + 2y + 5y$$
$$= 2x + 7y$$

$3x$ and $-x$ are like terms.
$+2y$ and $+5y$ are like terms.

$$7r + 3 - 2r - 8 = 7r - 2r + 3 - 8$$
$$= 5r - 5$$

$7r$ and $2r$ are like terms. The numbers are also terms, to be added or subtracted as normal.

Golden rules

- Each term in an expression includes the sign ($+$ or $-$) in front of it.
- If there is no sign, assume it is $+$.
- The term y means $1 \times y$. You do not need to write the 1.
- Like terms have **exactly the same** letter or letter combination.

Worked example

Simplify each expression.
(a) $a + a + a + a + a + a = 6a$
(b) $4f + 5f - 2f \qquad = 9f - 2f = 7f$
(c) $7q - 3p - q - 4p$
$= 7q - q - 3p - 4p = 6q - 7p$
(d) $3gh + 5h - 5gh - 8h$
$= 3gh - 5gh + 5h - 8h = -2gh - 3h$
(e) $3 + 5h - 5 - 8h$
$= 3 - 5 + 5h - 8h = -2 - 3h$

For part (c), rearrange to group the like terms:
$$7q - 3p - q - 4p = 7q - q - 3p - 4p$$

Like terms have **exactly the same** combination of letters. In part (d) $3gh$ and $5h$ are **not** like terms.

Problem solved!

- Label all the sides.
- The perimeter is the total distance around the shape, so write the perimeter as an addition.
- Simplify the expression.

You'll need brilliant problem-solving skills to succeed in GCSE – get practising now!

Worked example

Use algebra to work out the **perimeter** of each of these shapes.

(a)

$x + x + x + x$
$= 4x$

(b)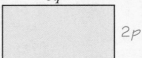

$2p + 5q + 2p + 5q$
$= 4p + 10q$

Now try this

1 Simplify each expression fully.

   (a) (i) $e + e + e + e$ (ii) $6z - 4z$

   (b) (i) $3j + 5k + 4j - 2k$ (ii) $-5 - 2a + 10 - 4a$

2 What is the perimeter of a rectangle with width $3x$ and length $4y$?

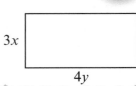

Simplifying expressions

You can simplify products by multiplying or dividing the individual components. You need to learn the rules so that you can simplify fully.

For a reminder about multiplying and dividing with negative numbers turn to page 10.

① Multiplying expressions

* Multiply the numbers together.
* Multiply the letters: $a \times b = ab$
* Use the rules for multiplying powers.

$a \times a = a^2$ •————• a squared
$b \times b \times b = b^3$ •————• b cubed

* Write the letters alphabetically.

$b \times a = ab$

$5a \times 3b = 15ab$
$5 \times 3 = 15 \quad a \times b = ab$

* The rules for multiplying with negative numbers still apply.

$4y \times -2y = -8y^2$
$4 \times -2 = -8 \qquad y \times y = y^2$

② Dividing expressions

* Write the division as a fraction.
* Cancel any number parts if possible.
* Use the rules for dividing powers to simplify any letter parts.
* The rules for dividing with negative numbers still apply.

$$-24d^3 \div 6d = -\frac{^{4}24d^{3\,2}}{_{1}6d}$$

$24 \div 6 = -4$

$$= -4d^2$$

$d^3 \div d = d^2$

Worked example

Simplify each expression.
(a) $t \times t$ $\qquad = t^2$
(b) $-2f \times 6e$ $\qquad = -12ef$
(c) $-3a \times -5a$ $\qquad = 15a^2$
(d) $2a \times -3b \times 4a$ $\qquad = -24a^2b$

(a) Write $t \times t$ as t^2 (t squared).
(b) Write the letters alphabetically.
(c) Use the rules for multiplying:
$- \times - = +$
(d) Multiply the numbers and multiply the letters: $2 \times -3 \times 4$ and $a \times a \times b$

For more on indices, turn to page 30.

(a) $(20 \div 4)y$
(b) Cancel $\frac{15}{5}$ and the letter k
(c) Use the rules for dividing positive by negative and for dividing powers.
$30 \div -6 = -5$ and $q^2 \div q = q$

For more on indices turn to page 30.

Worked example

Simplify each expression.
(a) $20y \div 4$ $\qquad = 5y$
(b) $15kn \div 5k$ $\qquad = 3n$
(c) $30q^2 \div -6q$ $\qquad = -5q$

Now try this

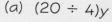

1 Simplify each expression fully.
(a) $-5a \times -4b$ (b) $6e \times -5f$

2 Simplify each expression fully.
(a) $18b \div 3$ (b) $40x^3 \div 8x$

3 Write an expression for the area of this rectangle. The measurements are in cm.

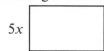

$5x$ [ ]
$4y$

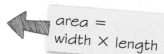
area = width × length

Writing expressions

You can write algebraic expressions to represent real-life quantities. These algebraic expressions represent the areas shaded green:

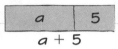

| a | 5 |

$a + 5$

| n | n |

$2n$

b
	3

$b - 3$

Golden rules

✓ Read the question carefully.

✓ Choose a suitable letter for each unknown quantity.

✓ Use the information you are given to build an expression to describe the situation, as in the example below.

$5b$

$4b$

The perimeter of this rectangle is given by the expression

$4b + 5b + 4b + 5b = 18b$

The area of the rectangle is given by the expression

$4b \times 5b = 4 \times 5 \times b \times b = 20b^2$

To revise the perimeter and area of rectangles, turn to pages 75 and 76.

Worked example

Match each expression with its meaning.

$n + 5$ half of a number
$\frac{n}{2}$ three times a number
$3n$ five more than a number
$n - 5$ double a number
$2n$ a number cubed
n^3 five less than a number

The letter n is used to represent a number. Think how you would express each calculation.

- Five more than n can be written as $n + 5$.
- Double n is $2 \times n = 2n$.
- Half of n is $n \div 2 = \frac{n}{2}$

For a reminder about cube numbers turn to page 12.

(a) When Max is 4, Oli will be $4 + 3$, so if Max's age is y then Oli's must be $y + 3$

(b) 5 bags of x apples gives $x + x + x + x + x = 5x$ apples, plus an extra 3 apples: $5x + 3$

Worked example

Write an expression to describe each situation.
(a) Max is y years old. Oli is 3 years older than Max. Write an expression for Oli's age.

$y + 3$

(b) A bag of apples contains x apples. Ruben buys 5 bags and 3 loose apples. Write an expression for the number of apples he buys.

$5x + 3$

After the first spoonful, $(200 - 5)$ ml are left, after the second, $200 - 2 \times 5$, and after the third, $200 - 3 \times 5$. What will be left after $(s \times 5)$ ml has been taken out?

Now try this

1 What is the **area** of a rectangle with length $5x$ and width $3y$?

2 Write an expression to describe the length of the grey bar.

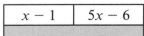

| $x - 1$ | $5x - 6$ |

3 A bottle contains 200 ml of medicine. A spoon holds 5 ml of medicine. Write an expression for the amount in ml that is left in the bottle after s spoonfuls have been poured out.

Indices

The rules for multiplying and dividing indices also apply in algebra.

For a reminder about indices turn to page 14.

base number ──a^{b}── index $a \times a \times a \times a \times a = a^5$

The rules of indices

1 When multiplying powers of the same base, add the indices.
$a^m \times a^n = a^{m+n}$
$y^4 \times y^3 = y^{4+3} = y^7$

2 When dividing powers of the same base, subtract the indices.
$a^m \div a^n = a^{m-n}$
$p^5 \div p^3 = p^{5-3} = p^2$

3 When raising a power to another power, mulitply the indices.
$(a^m)^n = a^{mn}$
$(t^4)^3 = t^{4 \times 3} = t^{12}$

4 Raising a base to a negative power is the same as the reciprocal of the base raised to a positive power.
$a^{-b} = \dfrac{1}{a^b}$
$c^{-3} = \dfrac{1}{c^3}$ ── If $c = 4$: $4^{-3} = \dfrac{1}{4^3} = \dfrac{1}{64}$

Watch out! You can only use the index laws when the bases are the same.

Worked example

Simplify
(a) $k^5 \times k^7$ $k^{5+7} = k^{12}$
(b) $g^6 \div g^2$ $g^{6-2} = g^4$
(c) $(h^2)^5$ $h^{2 \times 5} = h^{10}$
(d) $b^8 \times b \div b^3$ $b^{8+1-3} = b^6$
(e) $\dfrac{y^6 \times y}{y \times y^3}$ $= \dfrac{y^7}{y^4} = y^{7-4} = y^3$

For part (d) work out the multiplication first and then the division. Remember that $b = b^1$

For part (e) work out the top of the fraction and the bottom of the fraction separately, and then divide.

Worked example

(a) Write a^{-5} as a fraction. $\dfrac{1}{a^5}$
(b) Write $\dfrac{1}{b^{-2}}$ as a positive power. b^2
(c) Work out the value of 5^{-2} $\dfrac{1}{5^2} = \dfrac{1}{25}$

Worked example

Simplify fully.
(a) $3p^2 \times 2p^5$ $= 6p^7$
(b) $30q^6 \div 6q^3$ $= 5q^3$
(c) $-5y^2 \times -2y^3$ $= 10y^5$
(d) $-32a^5 \div 4a^3$ $= -8a^{5-3} = -8a^2$

(a), (b) Multiply the numbers and then multiply the powers.

(c) Remember to use the rules for multiplying negative numbers.

(d) Write as a fraction, cancel the number parts by dividing by 4, and apply the rules for powers.
$\dfrac{-^{8}\cancel{32}a^5}{_{1}\cancel{4}a^3} = -8a^{5-3}$

Look at page 10 to revise mulitplying and dividing with negative numbers.

Now try this

1 Simplify fully
   (a) $n^7 \times n^3$ (b) $m^4 \div m$ (c) $(k^3)^6$

2 Work out the value of 8^{-2}

3 Simplify fully
   (a) $4a^2 \times -3a^3$
   (b) $28m^4 \div 7m^2$
   (c) $5w^3 \times 4w^{-3}$

4 Work out $\dfrac{y^4 \div y}{y \times y^2}$

5 Use question 4 to show that $y^0 = 1$

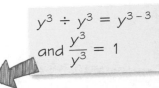
$y^3 \div y^3 = y^{3-3}$
and $\dfrac{y^3}{y^3} = 1$

Expanding brackets

You can use the **distributive law** to **multiply out** or **expand** brackets.

With numbers: $3 \times (4 + 5) = 3 \times 4 + 3 \times 5$

With algebra: $3(a + b) = 3a + 3b$

Golden rules

- Each term inside the brackets is multiplied by the term outside the brackets.
- You must take into account the sign before (to the left of) each term.

$3y \times 2y = 6y^2$

$3y(2y - 4x) = 6y^2 - 12xy$

$3y \times -4x$

- Take extra care when the term outside the brackets is negative.

For a reminder about multiplying with negative numbers turn to page 10 and to revise multiplying expressions turn to page 28.

For a reminder about multiplying with negative numbers turn to page 10 and to revise multiplying expressions turn to page 28.

Worked example

Expand the brackets.

(a) $6(2n - 3)$
$6 \times 2n + 6 \times -3 = 12n - 18$
(b) $3(4p + 7)$
$3 \times 4p + 3 \times 7 = 12p + 21$
(c) $t(4q - 3t)$
$t \times 4q + t \times 3t$
(d) $-5a(3b - 2a)$
$-5a \times 3b - 5a \times -2a = -15ab + 10a^2$

(a) Multiply $2n$ and -3 by 6
(c) Letters multiplied together should be in alphabetical order so write qt not tq.
(d) $- \times - = +$ so $-5a \times -2a = 10a^2$

Expand each set of brackets separately. Collect any like terms.

For a reminder about collecting like terms turn to page 27.

For a reminder about collecting like terms turn to page 27.

Be very careful with negative signs.

Worked example

Expand and simplify.

(a) $4(3b - 2a) + 5(4a + 3b)$
$12b - 8a + 20a + 15b$
$= 12b + 15b + 20a - 8a = 27b + 12a$
(b) $2x(4x - 6y) - 5y(3x - 4y)$
$8x^2 - 12xy - 15xy + 20y^2$
$= 8x^2 - 27xy + 20y^2$

Worked example

Show that $3a(4b - 2a) = -6a(a - 2b)$
LHS: $3a(4b - 2a) = 12ab - 6a^2$
RHS: $-6a(a - 2b) = -6a^2 + 12ab$
$= 12ab - 6a^2$
LHS = RHS ✓

Problem solved!

Expand the left-hand expression and then expand the right-hand expression to show that they both give the same terms.

You'll need brilliant problem-solving skills to succeed in GCSE – get practising now!

Now try this

1 Expand these expressions
   (a) $4(h - 5g)$ (b) $c(5d + 6)$

2 Expand $5x(3x - 4y + 8)$

3 Expand and simplify
   $5a(3b + 2a) + a(4a - b)$

4 Prove that $10e(e - 2f) = -5e(4f - 2e)$

Expanding double brackets

To expand **double** brackets (multiply them out), you can use the **grid method** or the **FOIL method**. In both cases, each term in one set of brackets is multiplied by each term in the other set.

1 The grid method

- Partition the terms in each bracket, including their signs, and write the terms as headings in a grid.
- Multiply the terms, writing each product in the correct cell.
- Collect any like terms.

To expand $(x + 3)(x - 5)$:

Partition the terms

	x	$+3$
x	x^2	$+3x$
-5	$-5x$	-15

Multiply

$(x + 3)(x - 5) = x^2 + 3x - 5x - 15$
$= x^2 - 2x - 15$

Collect like terms

2 The FOIL method

FOIL tells you the order in which to multiply the terms when you expand double brackets.

$$y^2 \quad -12$$

$$(y - 4)(y + 3) = y^2 + 3y - 4y - 12$$
$$= y^2 - y - 12$$
$$-4y$$
$$3y$$

collect like terms

First terms
Outer terms
Inner terms
Last terms

Worked example

Expand and simplify
(a) $(x + 5)(x + 4)$
$= x^2 + 4x + 5x + 20 = x^2 + 9x + 20$
(b) $(x + 2)(x - 8)$
$= x^2 - 8x + 2x - 16 = x^2 - 6x - 16$
(c) $(x - 5)(x - 6)$
$= x^2 - 6x - 5x + 30 = x^2 - 11x + 30$
(d) $(x + 4)(x - 4)$
$= x^2 - 4x + 4x - 16 = x^2 - 16$
(e) $(x - 9)^2$
$= (x - 9)(x - 9) = x^2 - 9x - 9x + 81$
$= x^2 - 18x + 81$

You can use either the grid method or the FOIL method. In each case you will get **two** x-terms, which you need to simplify.

In part (e) write $(x - 9)^2$ as $(x - 9)(x - 9)$ then multiply out.

Worked example

Without expanding, match each pair of brackets with its expansion.

$(x + 3)(x - 4)$ $x^2 - 7x + 12$
$(x + 3)(x + 4)$ $x^2 - 8x + 16$
$(x - 3)(x - 4)$ $x^2 - x - 12$
$(x + 3)(x - 3)$ $x^2 + 7x + 12$
$(x - 4)^2$ $x^2 - 9$

Problem solved!

Think about the number terms and their signs. The number term when $(x + 3)(x - 4)$ is expanded is
$3 \times -4 = -12$
You can use the x-term for matching if the number terms are the same. The x-term in $(x - 3)(x - 4)$ is $-4x - 3x = -7x$

You'll need brilliant problem-solving skills to succeed in GCSE – get practising now!

Now try this

1 Expand and simplify
   (a) $(x - 6)(x + 2)$ (b) $(x - 5)(x - 8)$
   (c) $(x - 5)(x + 5)$ (d) $(x + 4)^2$

2 Work out the expansion to $(x - 2)(x + 2)$ **without** using a written method.

Factorising

Factorising is the **inverse** or **opposite** of expanding brackets.

expanding

$$4y(3x - 5) = 12xy - 20y$$

factorising

Number factors

* Look for the HCF of the number parts of all the terms.

$12a - 18b$ ⟶ HCF of 12 and 18 is 6

* Write the HCF outside the brackets. Keep the sign the same inside the brackets.

$6(\quad - \quad)$

* Divide each term by the number outside the brackets to decide what to write inside.

$6(2a - 3b)$

$12a \div 6 = 2a \qquad 18 \div 6 = 3b$

Letter factors

* Look for letters that are common to all the terms.

$5xy - 3yz$ y is common $p^5 - 3p^2$ p^2 is common

* Write the common letter outside the brackets, dividing it into the terms inside.

$y(5x - 3z) \qquad p^2(p^3 - 3)$ $p^3 = p^5 \div p^2$

* **Check** by expanding the brackets again. You should get back to the original expression.

$y(5x - 3z) = 5xy - 3yz$ ✓

$p^2(p^3 - 3) = p^5 - 3p^2$ ✓

Worked example

Factorise completely
(a) $24xy - 16x$ $= 8x(3y - 2)$
(b) $12y^5 - 30y^3$ $= 6y^3(2y^2 - 5)$

Worked example

Factorise
(a) $3x + 12$ $= 3(x + 4)$
(b) $3ab - 4b$ $= b(3a - 4)$
(c) $4x^3 - 5x^2$ $= x^2(4x - 5)$

(a) 3 is a factor of both 3 and 12.
(b) b is a factor of $3ab$ and of $-4b$.
(c) The highest power of x in both terms is x^2.

Find the HCF of the number parts and of the letter parts.
$8x$ is the HCF of $24xy$ and $16x$.
$6y^3$ is the HCF of $12y^5$ and $30y^3$.
Write the HCF outside the brackets. Divide the terms inside the brackets by the HCF.
Check it!
(a) $8x \times 3y - 8x \times 2 = 24xy - 16$ ✓
(b) $6y^3 \times 2y^2 - 6y^3 \times 5 = 12y^5 - 30y^3$ ✓

Factors combining numbers and letters

* Look at all of the terms and check for common factors that include numbers **and** letters.

4 is the HCF of 8 and 12

$8x^2 - 12xy$ $4x$ is the HCF of $8x^2$ and $12xy$

x is the HCF of x^2 and xy

* Write the HCF outside the brackets and divide each term inside the brackets by it.

$8x^2 \div 4x = \dfrac{^2 8x^2 1}{^1 4x}$ $12xy \div 4x = \dfrac{^3 12xy y}{^1 4x ^1}$

$= 2x$ $= 3y$

Now try this

1 Factorise
  (a) $18y - 27q$ (b) $7pq - 5pt$
2 Factorise completely $21d^3 - 35d^5$

'Factorise completely' means taking the HCF outside the brackets. Look for a common number **and** letter factor.

Substitution

You work out the value of an expression by **substituting** or replacing the letters with their numerical value, remembering the priority of operations.

For a reminder about the priority of operations turn to page 13.

When $x = 5$ and $y = 3$:

$$4x^2 - 2y = 4 \times 25 - 2 \times 3$$
$$= 100 - 6$$
$$= 94$$

When $x = 5$,
$x^2 = 5 \times 5 = 25$

Indices first ($5^2 = 25$), then multiplication (4×25 and 2×3). Finally, do subtraction.

Worked example

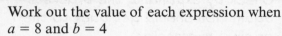

Work out the value of each expression when $a = 8$ and $b = 4$

(a) $a - b$ $8 - 4 = 4$
(b) $a(10 - 2b)$ $8(10 - 2 \times 4)$
$= 8(10 - 8) = 8 \times 2 = 16$
(c) $\dfrac{(a + b)^2}{b}$ $\dfrac{(8 + 4)^2}{4} = \dfrac{12^2}{4} = \dfrac{144}{4} = 36$

Substitute $a = 8$ and $b = 4$ into each expression. Use BIDMAS to make sure you calculate the answer using the correct priority of operations.

In part (c), work out the brackets first, then the indices (squared) and then divide.

Problem solved!

The easiest way to work out whether Freddie is correct is to work out the value of $3x^2$ when $x = 5$. If you show your working you will have explained your answer.

You'll need brilliant problem-solving skills to succeed in GCSE – get practising now! 💡

Worked example

Freddie says that the value of $3x^2$ when $x = 5$ is 225. Is Freddie correct? Give a reason for your answer.

$3x^2 = 3 \times 5^2 = 3 \times 25 = 75$
Freddie is not correct.
(Freddie has worked out $(3x)^2$, which is different from $3x^2$.)

Always write out the calculation with any values substituted **before** doing any calculations. Then use the correct priority of operations. Be careful with negative numbers:

$(-2)^3 = -2 \times -2 \times -2 = -8$

Worked example

Using the values of the letters in the table, work out the value of each expression.

Letter	p	q	x	y	z
Value	10	-2	15	-5	0

(a) $5(p + q + y)$
$5(10 + (-2) + (-5)) = 5 \times 3 = 15$
(b) $\sqrt{p + x} + 4y$
$\sqrt{10 + 15} + 4 \times (-5)$
$= \sqrt{25} + 4 \times (-5)$
$= 5 + -20 = -15$
(c) $(qy + pz)^2 - q^3$
$(-2 \times (-5) + 10 \times 0)^2 - (-2)^3$
$= 10^2 - -8$
$= 100 + 8 = 108$
(d) $q(p - y)^2$
$-2(10 - -5)^2 = -2(15)^2$
$= -2 \times 225 = -450$

Now try this

1 Work out the value of $8x - 5y$
  when $x = 10$ and $y = 5$

2 Work out the value of $-5b^3$
  when $b = -2$

3 Work out $\sqrt{ab - c}$
  when $a = 5$, $b = 8$ and $c = 4$

Linear sequences

A **linear sequence** is a pattern of numbers where the difference between consecutive terms is **constant**. Linear sequences are also called **arithmetic** sequences.

Each number in a sequence is a **term**. The rule that gets from one term to the next is called the **term-to-term** rule. Here are some examples:

Sequence:	5, 10, 15, 20, ...	1, 11, 21, 31, ...	4, 2, 0, −2, ...
Term-to-term rule:	+5	+10	−2

Worked example

1, 5, 9, 13, ... is a linear sequence.
(a) Write the term-to-term rule for this sequence.
5 − 1 = 4 and 9 − 5 = 4
The term-to-term rule is + 4
(b) Write the next three terms of the sequence.
17, 21, 25
(c) Write the 10th term of the sequence.
29, 33, 37
The 10th term is 37
(d) Will 100 be in this sequence? Give a reason for your answer.
No, because all the terms are odd and 100 is even.

(a) You know the sequence is linear, so the rule will be + ☐ or − ☐

(b) Add 4 to find the next term in the sequence.

(c) Continue adding 4 until you get the 10th number in the sequence (the 10th term).

(d) Look for patterns in the sequence. All the terms you have written down are odd numbers, so 100 cannot be a term in the sequence.

Work out what you need to add or subtract to get from each term to the next. If you add or subtract a constant amount then the sequence is linear.

Just because a sequence is not linear, it doesn't mean that it does not follow a pattern. Part (b) shows the sequence of **square numbers**. In part (c) the term-to-term rule is × 2. This is not a linear sequence but you could use the rule to continue the sequence.

Turn to page 37 for more about non-linear sequences.

Worked example

Which of these sequences are linear sequences? Give reasons for your answers.
(a) 2, 3, 4, 5, ...
3 − 2 = 1, 4 − 3 = 1, 5 − 4 = 1
Linear. The term-to-term rule is + 1
(b) 1, 4, 9, 16, ...
4 − 1 = 3, 9 − 4 = 5, 16 − 9 = 7
Not linear. The difference between terms is not constant.
(c) 2, 4, 8, 16, ...
4 − 2 = 2, 8 − 4 = 4, 16 − 8 = 8
Not linear. The difference between terms is not constant.
(d) 50, 40, 30, ...
40 − 50 = − 10, 30 − 40 = − 10
Linear. The term-to-term rule is − 10

Now try this

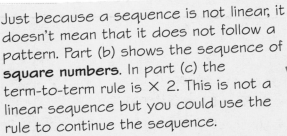

Here is a number sequence: 68, 59, 50, 41, ...

(a) Write the first term of the sequence.

(b) Write the term-to-term rule.

(c) Find the next three terms after 41.

(d) Is this sequence an arithmetic sequence?

(e) What is the first negative number in this sequence?

Any linear sequence with a 'subtract' rule will eventually produce negative terms. Keep subtracting the difference until you get a number less than 0.

The nth term

The **nth term** of a sequence is a rule for finding any term in the sequence. It is also called the **general term**. n is the position or term number, so n is always a positive whole number. To work out the terms of a sequence substitute $n = 1$ (1st term), $n = 2$ (2nd term), ... into the nth term.

nth term   sequence   nth term   sequence

$$2n - 1 \rightarrow 1, 3, 5, 7, \ldots \qquad 2 - 3n \rightarrow -1, -4, -7, -10, \ldots$$

$$2 \times 1 - 1 = 1 \quad 2 \times 2 - 1 = 3 \qquad 2 - 3 \times 1 = -1 \quad 2 - 3 \times 2 = -4$$

Worked example

(a) Find the fifth term in the sequence with nth term $3n + 2$

$3 \times 5 + 2 = 15 + 2 = 17$

(b) Work out the first three terms in the sequence with nth term $5n - 5$

$n = 1: 5 \times 1 - 5 = 5 - 5 = 0$
$n = 2: 5 \times 2 - 5 = 5$
$n = 3: 5 \times 3 - 5 = 10$

(c) What is the sixth term in the sequence with nth term $10 - 2n$?

$10 - 2 \times 6 = 10 - 12 = -2$

(d) Is 25 in the sequence with nth term $2n + 2$? Explain your answer.

No, $2n + 2$ terms are all even.

In parts (a)–(c) substitute the term number into the nth term.

In part (d) multiples of 2 ($2n$) are all even, and adding 2 to an even number will always result in an even number.

Golden rules

Follow these steps to work out the nth term of a sequence.

1 Work out the **common difference** between the terms. This is the coefficient (multiple) of n.

     +3   +3   +3

2,   5,   8,   11, ... $3n$

2 Use the inverse, working backwards, to find the 'zero' term. Subract the difference.

   −3   +3   +3   +3

−1,   2,   5,   8,   11, ...   −1

3 Combine both parts to get the nth term.

$3n - 1$

4 Check: substitute $n = 1$ and $n = 2$ into the nth term.

$n = 1: 3(1) - 1 = 2$ ✓
$n = 2: 3(2) - 1 = 5$ ✓

Worked example

Work out the nth term of each sequence.
(a) 4, 7, 10, 13, ...
difference is 3, so $3n$
'zero' term $4 - 3 = 1$, so rule is $3n + 1$
(b) 3, 8, 13, 18, 23 ...
difference is 5, so $5n$
'zero' term: $3 - 5 = -2$, so rule is $5n - 2$
(c) 5, 3, 1, −1 ...
difference is −2, so $-2n$
'zero' term: $5 - -2 = 7$, so rule is $-2n + 7$

For each sequence:
- work out the difference
- subtract the difference from the first term to find the zero term
- combine the zero term and the difference to give the nth term.

Now try this

1 Write the first five terms of the sequence with nth term $4n - 3$

2 Work out the nth term of each sequence.
(a) 3, 7, 11, 15, ...   (b) 4, 2, 0, −2, ...

Non-linear sequences

The terms in **non-linear** sequences increase or decrease by different step sizes.

Geometric sequences increase or decrease using multiplication or division to get from one term to the next. **Quadratic** sequences involve the term n^2.

$$2^n = 2^1, 2^2, 2^3, \ldots$$
$$\quad = 2, 4, 8, \ldots$$

$$n^2 = 1^2, 2^2, 3^2, \ldots$$
$$\quad = 1, 4, 9, \ldots$$

Geometric sequence: powers of 2   Quadratic sequence: square numbers

---

Worked example

Work out the next three terms in each sequence and explain the term-to-term rule.

(a) 1, 4, 9, 16, ...

Term-to-term rule: increase the difference by 2 each time   25, 36, 49

(b) 1, 2, 4, 8, ...

Term-to-term rule: double   16, 32, 64

(c) 1000, 100, 10, ...

Term-to-term rule: divide by 10

1, 0.1, 0.01

> Part (a) is the sequence of square numbers. You can work out the next term by adding 3, then 5, then 7, ... It is the **quadratic** sequence n^2.
>
> In part (b) each term is double the previous term. It is also the sequence of the powers of 2, 2^n.
>
> The sequences in parts (b) and (c) are all **geometric** sequences. They increase or decrease by multiplication or division.

Worked example

Write down the first five terms of each sequence.

(a) $2n^2$

$n = 1: 2 \times 1^2 = 2 \times 1 = 2$

$n = 2: 2 \times 2^2 = 2 \times 4 = 8$

$n = 3: 2 \times 3^2 = 2 \times 9 = 18$

$n = 4: 2 \times 4^2 = 2 \times 16 = 32$

$n = 5: 2 \times 5^2 = 2 \times 25 = 50$

2, 8, 18, 32, 50

(b) $n^2 - 1$

$n = 1: 1 - 1 = 0$ $n = 2: 4 - 1 = 3$

$n = 3: 9 - 1 = 8$ $n = 4: 16 - 1 = 15$

$n = 5: 25 - 1 = 24$

0, 3, 8, 15, 24

Worked example

Work out the nth term of these sequences by comparing them to n^2.

(a) 3, 6, 11, 18, ...

1	4	9	16	
3	6	11	18	$\Big) + 2$

$n^2 + 2$

(b) 10, 40, 90, 160, ...

1	4	9	16	
10	40	90	160	$\Big) \times 10$

$\quad\quad 10n^2$

> Write out the first four terms of n^2.
>
> Compare each new sequence, term by term, with the terms of n^2. What do you need to do to 1, 4, 9, 16, ... to get the terms in each sequence?

Now try this

1 Write the first five terms of the sequence $n^2 + 1$

2 Work out the nth term of the sequence 6, 9, 14, 21, ... by comparing it to the sequence n^2.

3 Sort these sequences into arithmetic, geometric or quadratic sequences.

   4, 8, 16, 32, ...   100, 91, 82, 73, ...

   0, 3, 8, 15, 24, ...   2, 13, 24, 35, ...

   25, 36, 49, 64, ...   1, 5, 25, 125, ...

> What has been added to 1, 4, 9, ... to get this new sequence?

> Arithmetic sequences increase or decrease by adding or subtracting the same amount. Geometric sequences increase or decrease by multiplying or dividing by the same amount each time. Quadratic sequences are related to the sequence n^2.

Writing equations

An **equation** is a mathematical sentence that tells you that the quantities on either side of the = sign are equal. You use a letter to represent an unknown quantity or **variable**. You can use an equation to describe a word problem. The equation $3x - 6 = 12$ has one variable (x). It tells you that if you multiply x by 3 and then subtract 6, the answer is 12.

$$3x - 6 = 12$$

Golden rules

✓ Read the question carefully and choose a letter to represent the unknown.

✓ Write an expression that describes the situation.

   Three times a number subtract one → $3n - 1$

✓ Use the information in the question to put your expression equal to a known number.

   Three times a number subtract one equals 29 → $3n - 1 = 29$

For a reminder on writing expressions turn to page 29.

Worked example

Posters cost £3 each plus a one-off charge for postage of £2. Max spends £17 buying posters. Write an equation for this word problem.

Cost of posters is $3 \times p = 3p$
Add the postage charge of £2 → $3p + 2$
$3p + 2 = 17$

Choose a letter to represent the unknown (p).
Write an expression for the cost of the posters.

Add the postage charge. Make your expression equal to the amount Max spends.

Worked example

The perimeter of a regular hexagon is 30 cm. Write an equation for this word problem using s for the length of one side.

$s + s + s + s + s + s = 6s$
$6s = 30$

Worked example

Write each statement as an equation. Use n for the unknown number.
(a) Half of a number is eight.
   $\frac{n}{2} = 8$
(b) Double a number, add seven, is eleven.
   $2n + 7 = 11$
(c) Six times a number, subtract nine, is three.
   $6n - 9 = 3$
(d) The square of a number is 144
   $n^2 = 144$

Problem solved!

A regular hexagon has six equal sides. The perimeter is the distance around the hexagon, so is the sum of all the sides. Use the information that the perimeter = 30 cm to write the equation.

You'll need brilliant problem-solving skills to succeed in GCSE – get practising now! 💡

Now try this

- Use real numbers or empty boxes if you are not sure.
- Half a number is written as a fraction with denominator 2.
- Double a number is 2 × the number, written as $2n$.
- The square of a number is written as n^2.

Write an equation for each word problem. Use the letter in brackets for the unknown number.
(a) Five times a number (n) subtract seven is eighteen.
(b) The perimeter of a regular octagon of side length (s) is 32 cm.
(c) My family bought cinema tickets (t) at £9 each, and there was a separate booking fee of £1. The total cost was £64.

Solving simple equations

Solving an equation means working out the value of the **unknown**, which is usually a letter. To do this, you must always do the **same operation** to **both sides** of the equation until you get the letter on its own.

Use inverse operations to solve equations:

- the inverse of addition is subtraction
- the inverse of multiplication is division.

$3x = 12 \quad (\div 3)$ ⟵ Write down the operation you are doing.

$x = 4$ $3 \times x = 12$, so use $\div$ (the inverse of x) to get x.

⟵ This is the solution to the equation.

 Worked example

Solve

(a) $a + 10 = 15 \qquad (-10)$

$a + 10 - 10 = 15 - 10$

$a = 5$

(b) $b - 15 = 35 \qquad (+15)$

$b - 15 + 15 = 35 + 15$

$b = 50$

(c) $20 - y = 12 \qquad (+y)$

$20 - y + y = 12 + y$

$20 = 12 + y \qquad (-12)$

$y = 20 - 12 = 8$

When using inverse operations, remember to do the same thing to both sides. Replace the unknown with a ☐ and work out the missing number to make the number statements true.

$10 + \square = 15; \quad \square - 15 = 35; \quad 20 - \square = 12$

Worked example

Solve

(a) $3q = 27 \qquad (\div 3)$

$q = 27 \div 3 = 9$

(b) $\frac{t}{2} = 20 \qquad (\times 2)$

$t = 20 \times 2 = 40$

(c) $x^2 = 36 \qquad (\sqrt{\ })$

$x = \sqrt{36} = 6$

 (c) $\square^2 = 36$ so work out the **square root** of 36.

Using inverse operations to solve two-step problems

Sometimes you need to do more than one step to solve an equation. It is important to write your working neatly.

Start a new line for each step. ⟶

$2x - 5 = 11 \qquad (+5)$ ⟵ Write down the operation you are doing.

$2x - 5 + 5 = 11 + 5$

Every line should have an = sign in it.

$2x = 16 \qquad (\div 2)$ ⟵ Do one operation at a time.

$2x \div 2 = 16 \div 2$

$x = 8$ ⟵ The solution to the equation.

Check it!

$2 \times 8 - 5 = 16 - 5 = 11 ✓$ ⟵ Substitute $x = 8$ into the equation to check it works.

 Worked example

Solve

(a) $6x - 2 = 28 \qquad (+2)$

$6x = 30 \qquad (\div 6)$

$x = 30 \div 6 = 5$

(b) $\frac{y}{3} + 5 = 12 \qquad (-5)$

$\frac{y}{3} = 7 \qquad (\times 3)$

$y = 3 \times 7 = 21$

(c) $\frac{q}{5} - 2 = 18 \qquad (+2)$

$\frac{q}{5} = 20 \qquad (\times 5)$

$q = 20 \times 5 = 100$

Now try this

1 Solve

   (a) $5x = 35$ (b) $21 - p = 12$

   (c) $\frac{b}{5} = 25$ (d) $a - 14 = 26$

2 Solve

   (a) $5x + 2 = 42$ (b) $7y - 6 = 15$

   (c) $\frac{a}{4} + 5 = 10$ (d) $\frac{b}{7} - 2 = 2$

 Always do the same operation on both sides.

Solving harder equations

An equation may include brackets or have an unknown on both sides of the equals sign. One side may be a fraction. You need to be able to solve all of these types of equation.

Unknowns on both sides

Elliminate one of the terms: $+$, $-$, $\times$ or $\div$ to get only one unknown on one side of the $=$ sign.

To solve $7x + 3 = 3x - 9$

- Subtract $3x$ from both sides.

$4x + 3 = -9$

- Solve in the usual way.

$4x = -12$

$x = -3$

Brackets

Follow these steps to solve $3(2y + 7) = 3(y - 8)$

- Expand the brackets.

$6y + 21 = 3y - 24$

- Simplify by collecting like terms.

$3y = -45$

- Solve the equation.

$y = -15$

Fractions

Follow these steps to solve $\dfrac{3x - 4}{2} = x - 1$

- Multiply both sides by the denominator to get rid of the fraction.

$\dfrac{^1\cancel{2}(3x - 4)}{\cancel{2}_1} = 2(x - 1)$

- Expand the brackets, simplify and solve.

$3x - 4 = 2x - 2$

$x = -6$

Worked example

Solve
(a) $7x - 10 = 5x + 6$

$2x - 10 = 6$

$2x = 16$

$x = 8$

(b) $4(2y - 3) = 44$

$8y - 12 = 44$

$8y = 56$

$y = 7$

(c) $\dfrac{3n - 8}{2} = 5n - 3$

$3n - 8 = 2(5n + 3) = 10n + 6$

$3n - 14 = 10n$

$-14 = 7n$

$n = -2$

(a) Subtract $5x$ from both sides.
(b) Or you could divide both sides by 4 and cancel:

$\dfrac{^1\cancel{4}(2\ y - 3)}{_1\cancel{4}} = \dfrac{^{11}\cancel{44}}{_1\cancel{4}}$ giving $2y - 3 = 11$

Worked example

When I double my number and add 10, the answer is the same as when I treble my number and add 3. What is my number?

$2n + 10 = 3n + 3$ $(- 2n)$

$10 = n + 3$ $(- 3)$

$n = 7$

Double n, add 10, is $2n + 10$
Treble n, add 3, is $3n + 3$

Now try this

1 Solve

(a) $8x - 6 = 3x + 9$ (b) $5(4x - 6) = 90$

(c) $3(5y + 2) = 2(3y - 6)$ (d) $\dfrac{5 + 3x}{4} = 2x - 5$

Decide which steps you need to follow to get the letter term on its own.

2 Ruben and Finn start with the same number. Ruben trebles his number and adds 4. Finn multiplies his number by 5 then subtracts 6. They get the same answer. What number did they start with?

Write two expressions, one for Ruben's number and one for Finn's.

Inequalities

The symbols $<$, $>$, $\leqslant$ and $\geqslant$ describe **inequalities**. The open end of the symbol is always next to the bigger quantity. You can represent inequalities on a number line.

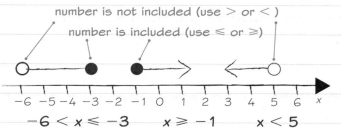

number is not included (use $>$ or $<$)

number is included (use $\leqslant$ or $\geqslant$)

$-6 < x \leqslant -3$ $x \geqslant -1$ $x < 5$

Integer solutions

These are whole numbers (positive, negative or zero) that **satisfy** the inequality.

- From a diagram or number line

Possible integer values are:
$-1, 0, 1, 2$

- From an inequality statement
  $-5 < x \leqslant -1$

Possible integer values are:
$-4, -3, -2, -1$

Worked example

(a) Show the inequality $-3 < x < 2$ on a number line.

(b) Write the inequality shown on the number line. List the integer values of x.

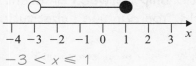

$-3 < x \leqslant 1$

-3 is not included, so the integer values are $-2, -1, 0, 1$

(a) The signs are $<$ and $<$ so -3 and 2 are not included – draw **empty** circles.

(b) The empty circle means -3 is not included (use $<$). The filled circle means 1 is included (use $\leqslant$).

Solving inequalities

You can solve an inequality in the same way as you solve an equation. You use an inequality sign instead of an $=$ sign.

$2x - 4 > 12$ $(+ 4)$
$\quad 2x > 16$ $(\div 2)$
$\quad\quad x > 8$

For a reminder about solving equations turn to pages 39 and 40.

Solve in the same way as you would an equation.

Worked example

Solve
(a) $3x - 4 \geqslant 8$ $(+ 4)$
$\quad 3x \geqslant 12$ (23)
$\quad\quad x \geqslant 4$
(b) $\frac{x}{5} 1 6 , 9$ (-6)
$\quad \frac{x}{5} < 3$ $(\times 5)$
$\quad\quad x < 15$

Now try this

1 Write the inequality shown on the number line.

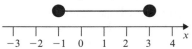

Solve both inequalities to see if any values for x are solutions for **both** inequalities.

2 Write the integer values of x that satisfy $-2 \leqslant x < 4$

3 Draw the inequality $x > -1$ on a number line.

4 Solve the inequality $5x + 7 < 22$

5 Can the inequalities $8 - 2x \leqslant 10$ and $3x - 4 < 2$ both be true at the same time? Explain why.

Expression, equation, identity or formula?

You need to know the difference between expressions, equations, identities and formulae.

$5x - 4 = 11$	$3a - b^2$	$a - b \equiv -b + a$	$C = \pi d$
equation	expression	identity	formula

Golden rules

- An **expression** contains numbers, variables (letters such as x or y) and operators ($+$, $-$, $\times$, $\div$); it has no $=$ sign so you cannot work out what the values of the letters are.

  $3x + 4y - 9$ is an expression. x and y can have many different values.

- An **equation** is made up of two expressions on either side of an equals sign. When there is only one variable you can work out its value.

  $3x + 5 = 26$ and $7y - 4 = 5y + 6$ are both equations. You can work out the value of x and the value of y by solving each equation.

- An **identity** is an equation that is true for **all** values of the variables. Use the $\equiv$ sign instead of $=$ when you know it is an identity.

  $a + b \equiv b + a$ and $xy \equiv yx$ are identities.

- A **formula** is a mathematical rule that shows the relationship between different variables. Formulae are often related to the real world. You can use substitution to work out unknown values.

  $V = l \times w \times h$ (or $V = lwh$) is the formula for the volume of a cuboid, where V stands for volume, l for length, w for width and h for height.

Worked example

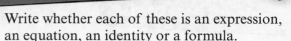

Write whether each of these is an expression, an equation, an identity or a formula.

$2x + 6 = 24$	$A = \pi r^2$	
equation	formula	
$5ab - c$	$5a - 6 = 3a + 8$	$4x^2 = (2x)^2$
expression	equation	identity
$b - a = -a + b$	$x^2 = 8x$	$S = \dfrac{D}{T}$
identity	equation	formula

- The one without an $=$ sign is an expression.
- Formulae show a relationship between two or more different variables.
- Equations can be solved to find the values of the variable(s).
- Identities will always be true, for **all** values of the variables.

Now try this

1 Write down whether each of these is an expression, an equation, an identity or a formula.
   (a) $3a - 4$
   (b) $3a - 4 = 26$
   (c) $3a - 4 = 2a + 2$
   (d) $A = \frac{1}{2}bh$
   (e) $3a(b - 2a + c)$
   (f) $3(a + b) = 3a + 3b$

2 Substitute three pairs of values for a and b to show that $(a + b)(a - b) \equiv a^2 - b^2$

Using $a = 5$ and $b = 2$ gives
$(5 + 2)(5 - 2) = 7 \times 3 = 21$
and $5^2 - 2^2 = 25 - 4 = 21$ as well.
Now try two more pairs of values.

Formulae

A **formula** is a mathematical rule showing the relationship between two or more quantities. The plural form is **formulae**. The **subject** of the formula is the variable on its own before the = sign.

These are all formulae: $A = \pi r^2$ $C = \pi d$ $S = 2(ab + bc + ac)$

subjects of the formulae

Rearranging a formula

In the formula $F = ma$, F is force in newtons, m is mass in kg and a is acceleration in m/s².

You can rearrange the formula to make a different variable the subject.

Rearranging $F = ma$ gives $a = F \div m = \dfrac{F}{m}$ and $m = F \div a = \dfrac{F}{a}$

a is the subject m is the subject

Worked example

The formula connecting speed (S), distance (D) and time (T) is $D = ST$

(a) Work out D when $S = 50$ and $T = 4$

$D = 50 \times 4 = 200$

(b) Work out S when $D = 140$ and $T = 2$

$D = ST$
$140 = S \times 2$
$S = 70$

(a) Substitute 50 for S and 4 for T into the formula.

(b) You can substitute the values given for D and T and solve for S, or you can rearrange the formula to make S the subject first:
$D = ST$ so $S = D \div T$, then substitute.

Worked example

Rearrange the formula $P = 3t - 5$ to make t the subject.

$P = 3t - 5$ $(+5)$
$P + 5 = 3t$ $(\div 3)$
$t = \dfrac{P+5}{3}$

In part (a) substitute 77 for F and solve for C.

For a reminder about cancelling fractions look at page 22.

In part (b) use the same strategies as for solving equations.

To revise solving equations look at pages 39 and 40.

Subtract y from both sides of the equation and divide both sides by 2.

Worked example

The formula for changing between the Celsius and Fahrenheit temperature scales is
$$C = \tfrac{5}{9}(F - 32)$$
where C is the temperature in Celsius and F is the temperature in Fahrenheit.

(a) Change 77°F to degrees Celsius.
$C = \tfrac{5}{9}(77 - 32) = \tfrac{5}{\cancel{9}} \times {}^{5}\cancel{45}$
$\qquad\qquad\qquad = 25°C$

(b) Change 100°C to degrees Fahrenheit.
$100 = \tfrac{5}{9}(F - 32)$ $(\times 9)$
$900 = 5(F - 32)$ $(\div 5)$
$180 = F - 32$ $(+ 32)$
$F = 212°F$

Now try this

1 The formula for the cost, C, in pounds, of hiring a boat for h hours is $C = 15 + 3h$
   (a) What is the cost of hiring a boat for 4 hours?
   (b) Tom pays £33. How many hours did he hire the boat for?

2 Rearrange $d = 2x + y$ so that x is the subject of the formula.

Writing formulae

Writing a formula to solve a problem is a very useful skill.

A cleaner charges £5 travelling expenses plus £9 per hour.
(a) Write a formula to describe her total charge, £C, for h hours' work.
$C = 9h + 5$
(b) What is her charge for 5 hours' work?
$C = 9 \times 5 + 5 = 45 + 5 = 50$ £50
(c) She charges Annie £32 for some cleaning. How many hours of cleaning did Annie pay for?
$32 = 9h + 5$
$9h = 27$
$h = 3$ 3 hours

Golden rules

Follow these steps when writing a formula to solve a problem.
✓ Use a different letter to represent each quantity. Make the letter meaningful, such as C for charge.
✓ Substitute all the values you are given.
✓ Solve the equation.
✓ You may need to rearrange the equation or formula to make a different letter the subject.

(a) You want to find a formula for C so begin your formula $C = ...$
One hour costs £9, so the cost for h hours is 9h. Then add the £5 travelling charge to get the final formula.
(b) Substitute 5 for h into the formula.
(c) Substitute 32 for C and solve for h.

Or, in part (c) you could rearrange the formula to make A the subject before you substitute:
$T = 8A + 5C$
$T - 5C = 8A$
$A = \dfrac{T - 5C}{8}$

Worked example

A fitness class costs £5 for children and £8 for adults.
(a) Write a formula for the total cost (T) for A adults and C children.
Each adult costs £8, so A adults cost 8A
Each child costs £5 so C children cost 5C
$T = 8A + 5C$
(b) What is the cost for 3 adults and 4 children?
$T = 8 \times 3 + 5 \times 4 = 24 + 20 = 44$
Cost is £44.
(c) The total cost for 6 children and some adults is £46. How many adults went with the children?
$46 = 8A + 5 \times 6 = 8A + 30$
$46 - 30 = 8A$
$A = 16 \div 8 = 2$
2 adults went.

Now try this

1 A waiter earns £7 an hour plus tips.
(a) Write a formula for his total wage W for h hours' work plus t in tips.
(b) How much does he earn for 5 hours' work with £12.50 in tips?
(c) The next day he worked for 8 hours and earned £67.25 including tips. How much in tips did he earn?
(d) Over the weekend he earned £99, which included £15 in tips. How many hours did he work?

2 (a) Write a formula for the perimeter, P, of a regular 15-sided polygon with sides of length s.
(b) What is the perimeter if the sides are length 10 cm?
(c) The perimeter is 75 cm. What is the length of one of the sides?

The perimeter is the total distance around the shape, so it is the sum of all the sides.

Coordinates and midpoints

Coordinates describe the **position** of a point on a grid.

You always write coordinates in brackets, like this:

(x-coordinate, y-coordinate).

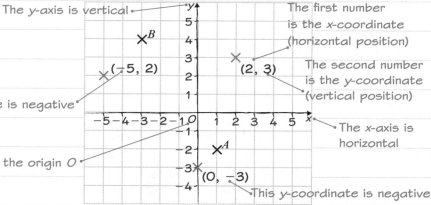

The y-axis is vertical

This x-coordinate is negative

The point (0, 0) is called the origin O

The first number is the x-coordinate (horizontal position)

The second number is the y-coordinate (vertical position)

The x-axis is horizontal

This y-coordinate is negative

Worked example

Look at the grid above. What are the coordinates of

(a) point A   (1, −2)
(b) point B?   (−3, 4)

Move across the x-axis to find the x-coordinate.

Move up or down the y-axis to find the y-coordinate.

Golden rules

To find the midpoint of a line segment without drawing a grid:

✓ add the x-coordinates and divide by 2

✓ add the y-coordinates and divide by 2.

$$\left(\frac{x_1 + x_2}{2}, \frac{y_1 + y_2}{2}\right)$$

Midpoint

The midpoint of a line segment is exactly halfway along the line. You can find the midpoint if you know the coordinates of the points at the ends of the line.

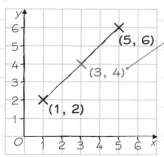

The midpoint is the point exactly halfway along the line

$$\text{Midpoint} = \left(\frac{1+5}{2}, \frac{2+6}{2}\right)$$
$$= \left(\frac{6}{2}, \frac{8}{2}\right) = (3, 4)$$

Worked example

Find the midpoint of the line joining the points (2, 3) and (8, 5).

$$\text{Midpoint} = \left(\frac{2+8}{2}, \frac{3+5}{2}\right) = \left(\frac{10}{2}, \frac{8}{2}\right) = (5, 4)$$

Now try this

1   (a)   Write the coordinates of
     (i)   point A
     (ii)   point B.
   (b)   On the grid, plot the point C so that ABC is a right-angled triangle.

2   Find the coordinates of the midpoint of the line joining the points (2, 5) and (4, 1).

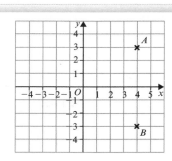

Gradient

The **gradient** of a straight-line graph tells you how steep the line is. To work out the gradient of a line you can draw a right-angled triangle. Work out the distance up and the distance across.

$$\text{Gradient} = \frac{\text{distance up}}{\text{distance across}}$$

The gradient can be **positive** or **negative**.

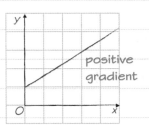
positive gradient

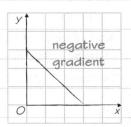
negative gradient

Worked example

Find the gradient of this line.

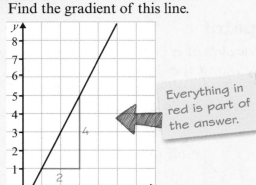

Everything in red is part of the answer.

1. Choose points on the line that have whole-number coordinates.
2. Draw a right-angled triangle.
3. Write the distance up and the distance across.
   Distance up = 5 − 1 = 4
   Distance across = 3 − 1 = 2
4. Use the formula for gradient.
5. The line slopes up from left to right so the gradient is positive.

$$\text{Gradient} = \frac{\text{distance up}}{\text{distance across}} = \frac{4}{2} = 2$$

Worked example

Draw a line with a gradient of $\frac{1}{3}$ on the grid below.

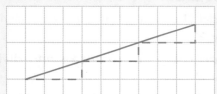

A gradient of $\frac{1}{3}$ means

$\frac{1}{3}$ — up distance, across distance

For every '3 squares across' move '1 square up'.

Now try this

Find the gradient of each line.

(a)

Gradients can be fractions.

(b)
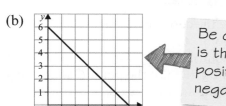

Be careful — is the gradient positive or negative?

$y = mx + c$

A straight-line graph has an equation of the form

$$y = mx + c$$

For a reminder about how to find the gradient of a line turn to page 46.

m is the gradient of the line — the gradient measures how steep the line is

c is the *y*-intercept, the point where the line crosses the *y*-axis

Worked example

Find the equation of this straight line.

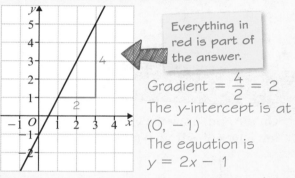

Everything in red is part of the answer.

Gradient $= \dfrac{4}{2} = 2$

The *y*-intercept is at $(0, -1)$

The equation is $y = 2x - 1$

Finding equations

To find the equation of the line you need to know the gradient and the *y*-intercept.

✓ Find the gradient, *m*.
  • Draw a right-angled triangle.
  • Gradient $= \dfrac{\text{distance up}}{\text{distance across}}$
✓ Find the *y*-intercept, *c*.
  Look at the point where the line crosses the *y*-axis.
✓ Write the equation.
  Put your values for the gradient, *m* and *y*-intercept, *c*, into the equation of a straight line,
  $y = mx + c$

Horizontal and vertical lines

• Horizontal lines have the equation
  $y = a$
• Vertical lines have the equation
  $x = a$

where *a* is a number.

Worked example

Write the letter of each graph line next to the correct equation.

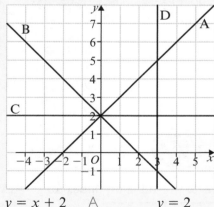

$y = x + 2$ A $y = 2$ C
$x = 3$ D $y = 2 - x$ B

Now try this

Write the letter of each graph line next to the correct equation.

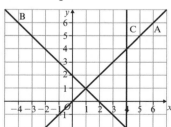

$y = -x + 2$

$x = 4$

$y = x$

Line A crosses the **y-axis** at **2** and has a **positive gradient**.
Line B crosses the **y-axis** at **2** and has a **negative gradient**.
Line C crosses the **y-axis** at **2** and has a **gradient of 0**.
Line D does not cross the **y-axis**.

Straight-line graphs

You can draw a straight-line graph by making a table of values and then plotting the points.

Worked example

Complete this table of values for the equation $y = 2x + 1$ for values of x from 0 to 5.

x	0	1	2	3	4	5
y	1	3	5	7	9	11

Notice the patterns in the table. The x-values increase by 1 and the y-values increase by 2 each time.

Substitute the x-value into the equation.
When $x = 0$: $y = (2 \times 0) + 1 = 1$
When $x = 1$: $y = (2 \times 1) + 1 = 2 + 1 = 3$
When $x = 2$: $y = (2 \times 2) + 1 = 4 + 1 = 5$
When $x = 3$: $y = (2 \times 3) + 1 = 6 + 1 = 7$
When $x = 4$: $y = (2 \times 4) + 1 = 8 + 1 = 9$
When $x = 5$: $y = (2 \times 5) + 1 = 10 + 1 = 11$

To plot a straight-line graph using a table of values, you must:
1. Make a table of values.
   The question tells you to plot the graph for values of x from 0 to 6 so your table should go from 0 to 6.
2. Substitute each value of x into the equation to find the y-value.
   When $x = 1$: $y = -\frac{1}{2} \times 1 + 3 = 2.5$
   so $y = 2.5$
3. Plot the points from the table on the grid.
   The first point is (0, 3).
4. Use a ruler to join the points with a straight line.

Worked example

On the grid, draw the graph of $y = -\frac{1}{2}x + 3$ for values of x from 0 to 6.

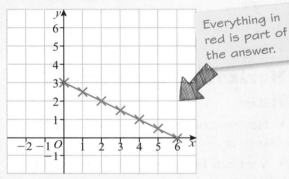

Everything in red is part of the answer.

x	0	1	2	3	4	5	6
y	3	2.5	2	1.5	1	0.5	0

Notice the patterns in the table. The x-values increase by 1 and the y-values decrease by 0.5

Now try this

(a) Complete this table of values for $y = 2x - 1$

x	0	1	2	3
y	−1			

(b) On the grid, draw the graph of $y = 2x - 1$

(c) Write the gradient of the line.

Compare the equation of the line with the general equation of a straight line, $y = mx + c$.

Formulae from graphs and tables

You can plot a real-life graph from a formula. You can then use the graph to solve problems.

Worked example

The cost of hiring a car is £40 plus £20 for each hire day.

(a) Write a formula for the cost, C, in pounds, of hiring a car for d days.

C = 40 + 20d

(b) Complete this table of values for the first 4 days of car rental.

Number of days, d	1	2	3	4
Cost, C (£)	60	80	100	120

(c) Plot a graph of this information on the axes below.

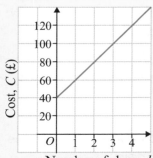

Notice the sequence in the table goes up by 20. This is the gradient of the graph. The y-intercept of the graph is the fixed charge for hiring the car.

Worked example

The graph below shows a plumber's charges for her work.

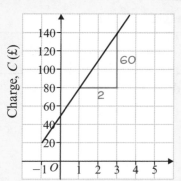

(a) Use the graph to complete this table of values.

Number of hours, h	0	1	2	3
Charges, C (£)	50	80	110	140

(b) How much does the plumber charge per hour?

$\frac{60}{2}$ = £30

(c) How much is the plumber's fixed charge?

£50

(d) Write a formula for the plumber's charges.

C = 30h + 50

In part (b) the plumber's charge per hour is given by the gradient. You can also look at how much the charge in the table increases every hour. In part (d) the formula is given by the hourly rate × number of hours worked plus the fixed charge.

Now try this

A builder charges £25 for each hour he works at a job, plus a £50 call-out fee.

(a) Write a formula for the amount he charges, C (in £), for h hours worked.

(b) Complete this table.

Number of hours, h	0	1	2	3	4
Charge, C (£)					

(c) Plot a graph of his charges against the hours he works. Draw your horizontal axis from 0 to 5 and your vertical axis from 0 to 200.  Your vertical scale should go up in 25s.

(d) For how many hours does he need to work on one job to earn £175?

Plotting quadratic graphs

An equation in which x^2 is the highest power of x is called a **quadratic equation**.

$$x^2 = 10 \qquad x^2 + 3x = 4 \qquad x^2 + 2x + 1 = 5$$

x^2 is the highest power so these are quadratic equations

You can draw the graph of a quadratic equation by completing a table of values.

Worked example

(a) Complete the table of values for $y = x^2 - 3x$

x	-1	0	1	2	3	4
y	4	0	-2	-2	0	4

(b) On the grid, draw the graph of $y = x^2 - 3x$

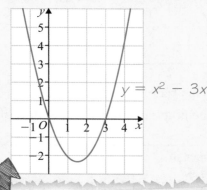

$y = x^2 - 3x$

Substitute each value of x into the equation to find the y-value.

When $x = -1$: $y = (-1)^2 - (3 \times -1)$
$\qquad = 1 - (-3) = 4$

When $x = 0$: $y = (0)^2 - (3 \times 0)$
$\qquad = 0 - 0 = 0$

When $x = 1$: $y = (1)^2 - (3 \times 1)$
$\qquad = 1 - 3 = -2$

When $x = 4$: $y = (4)^2 - (3 \times 4)$
$\qquad = 16 - 12 = 4$

Check that all the points on your graph lie on the curve. If a point doesn't lie on the curve double check the calculation.
Join the bottom two points with a smooth curve, **never** a straight line.

Drawing quadratics

✓ Make a table of values.

✓ Substitute the values of x to get the y-values.

✓ Plot the points.

✓ Draw a smooth curve that passes through every point.

✓ Label your graph.

Shape of quadratic curves

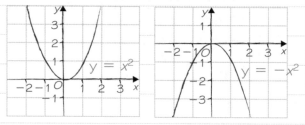

$y = x^2$ \qquad $y = -x^2$

Quadratic curves are always **curved** and **symmetrical**.

- When x^2 is positive the shape of the curve is like a smile ($\smile$).
- When x^2 is negative the shape of the curve is like a frown ($\frown$).

Now try this

(a) Complete the table of values for $y = x^2 - 1$

x	-2	-1	0	1	2
y	3			0	

(b) On a grid, draw the graph of $y = x^2 - 1$

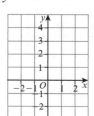

Be careful when substituting $x = -1$. Remember $(-1)^2 = 1$

Real-life graphs

Make sure you know how to use and interpret real-life graphs. Here are some examples:

Temperature conversion

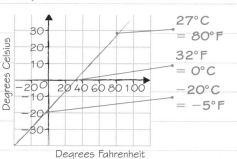

27°C = 80°F
32°F = 0°C
−20°C = −5°F

Conversion graph

Plumber's charges

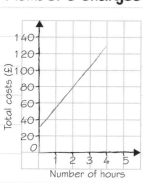

Initial fixed charge

This shows a plumber's charges of £25 per hour with an initial call-out fee of £30.

Average daily temperature

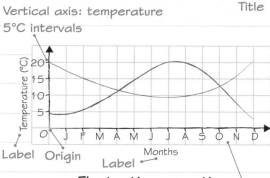

Vertical axis: temperature
5°C intervals

Title

Label   Origin   Label   Months

Fluctuation over time

Key — Australia   Horizontal axis: months
    — UK   no scale interval

Golden rules

✓ Read the title and labels on the axes.

✓ Look at both axes to work out the scale on each. The vertical scale may be different from the horizontal scale.

✓ For graphs showing more than one line, look at the key or explanatory labels.

(a) Product A is the only graph with a positive gradient.

(b) The intersection of the three lines shows when the products had similar sales.

(c) Read up from 2010 to the line for product C and then across to the sales axis. It is slightly more than halfway between £150 and £200, so it is about £180 000.

Worked example

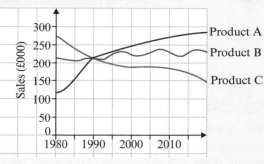

(a) Which product shows an increase in sales over the period 1980–2015? Explain how you can tell.

Product A. The line has a positive gradient.

(b) In which year were the sales similar for all products?

1990

(c) How much were the sales for product C in 2010?

Approximately £180 000

Now try this

Use this currency conversion graph to answer the questions.

(a) How many pounds will you get for $50?

(b) How many euros will you receive for £75?

(c) Hollie brings back $150 from the USA. She changes these into euros. How many euros will she get?

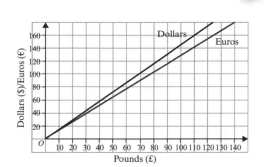

$150 = $100 + $50. Use the graph to change $150 to pounds and then change the pounds to euros.

51

Algebra problem-solving

You can use algebra to solve problems in other areas of maths.

Worked example

A chef uses a formula to calculate the length of time needed to cook a turkey. The cooking time t (in minutes) is 40 minutes for every kilogram plus an extra 50 minutes.

(a) Write a formula for the time t it takes to cook a turkey weighing w kg.

$t = 40w + 50$

(b) Use the formula to work out the time needed to cook a 4.5 kg turkey.

$t = 40 \times 4.5 + 50 = 230$ minutes
The time is 3 hours 50 minutes

(c) The chef cooks a turkey for 3 hours. What was its weight?

3 hours = 180 minutes
$180 = 40w + 50$
$40w = 130$
$w = 3.25$ kg

Problem solved!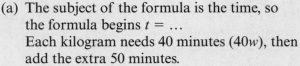

(a) The subject of the formula is the time, so the formula begins $t = \ldots$
Each kilogram needs 40 minutes ($40w$), then add the extra 50 minutes.

(b) Substitute $w = 4.5$ into the formula.

(c) First convert the time in hours to minutes. Then substitute the values you know and rearrange the formula to solve for w.

You'll need brilliant problem-solving skills to succeed in GCSE – get practising now!

Worked example

(a) Write an equation for the sum of the angles in this triangle.

$x + 2x + 3x = 6x$
$6x = 180°$

(b) Solve the equation to find the value of x.

$x = 180° \div 6 = 30°$

(c) Work out the size of each angle.

$x = 30°$
$2x = 2 \times 30° = 60°$
$3x = 3 \times 60° = 90°$
Angles are 30°, 60°, 90°

(d) What type of triangle is this?

One angle is 90°, so it is a right-angled triangle.

Problem solved!

(a) Add up all the angles. The angles in a triangle add to 180° so use this to form an equation.

(b) Divide both sides by 6.

(c) Substitute $x = 30°$ to find the other angles.

You'll need brilliant problem-solving skills to succeed in GCSE – get practising now!

Now try this

This graph converts approximately between ounces and grams.

(a) Convert 10 ounces to grams.
(b) Convert 125 grams to ounces.
(c) Convert 1 kg to pounds and ounces.
(d) 16 ounces (oz) = 1 pound (lb). Convert 1 lb 10 oz to grams.
(e) Explain why it is important to use either ounces or grams in a recipe and not a mixture of both.

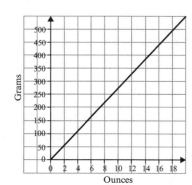

For part (c), the vertical scale doesn't go up to 1 kg, so pick a number on the scale that you can easily scale up to 1 kg.
1 kg = 1000 g = 2 × 500 g

Metric measures

Metric units are the units of measurement used in Europe and, mostly, in the UK.
You can convert between metric units by multiplying or dividing by 10, 100 or 1000.

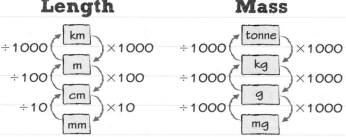

Length · **Mass** · **Capacity**

Worked example

(a) Convert 3 km to metres (m).

$$\times 1000$$

km $\longrightarrow$ m

$\times 3 \left(\begin{array}{cc} 1 & 1000 \\ 3 & 3000 \end{array} \right) \times 3$

$3 \times 1000 = 3000\,m$

(b) Convert 5000 m to kilometres (km).

$$\div 1000$$

m $\longrightarrow$ km

$\times 5 \left(\begin{array}{cc} 1000 & 1 \\ 3000 & 5 \end{array} \right) \times 5$

$5000 \div 1000 = 5\,km$

Draw a diagram to show the units you want to convert.
(a) From metres to km, ÷ 1000.
(b) From km to metres, × 1000.
Check it!
A metre is smaller than a kilometre, so the number will be larger. ✓

Word parts

'cent' means 100 – a century is 100 years

'kilo' is Greek and 'milli' is Latin; both mean 1000

Problem solved!

1. Work out the mass of the bag in grams.
2. The bag of marbles has a mass of 4000 g and there are 500 marbles in the bag, so divide the mass in grams by 500 to find the mass of one marble.

You'll need brilliant problem-solving skills to succeed in GCSE – get practising now! 💡

Worked example

The mass of a bag of marbles is 4 kg.
There are 500 marbles in each bag.
Work out the mass of one marble.

$$\times 1000$$

kg $\longrightarrow$ g

$\times 4 \left(\begin{array}{cc} 1 & 1000 \\ 4 & 4000 \end{array} \right) \times 4$

$4 \times 1000 = 4000$

$4000\,g \div 500 = 8\,g$

Now try this

1 Convert

(a) 300 cm to metres

(b) 8000 ml to litres

(c) 7 kg to grams

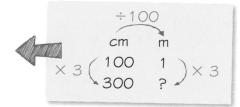

$$\div 100$$

cm $\longrightarrow$ m

$\times 3 \left(\begin{array}{cc} 100 & 1 \\ 300 & ? \end{array} \right) \times 3$

2 How many 200 ml glasses of milk can be filled from a 2-litre bottle of milk?

Work out how many ml there are in 2 litres.

Time

You need to be able to convert between different measurements of time. You also need to know how many days there are in each month.

Time conversions to learn

1 century = 100 years	1 week = 7 days	**Days in the months**
1 decade = 10 years	1 day = 24 hours	28 (29): Feb
1 year = 12 months = 52 weeks	1 hour = 60 minutes	30: Apr, June, Sept, Nov
1 (leap) year = 365 (366) days	1 minute = 60 seconds	31: Jan, Mar, May, July, Aug, Oct, Dec

Worked example

How many …
(a) decades in 50 years   5
(b) weeks in 42 days   6
(c) days altogether from 1 September to 31 December   122
(d) days in 5 years (include 1 leap year)

$4 \times 365 + 366 = 1826$

(e) seconds in $3\frac{1}{4}$ minutes?

$3.25 \times 60 = 195$

(a) $50 \div 10$
(b) $42 \div 7$
(c) $30 + 31 + 30 + 31$
(d) 1 year is 365 days
(e) 1 minute is 60 seconds; $\frac{1}{4} = 0.25$

Minutes as a fraction of 1 hour

1 minute is $\frac{1}{60}$ of an hour

5 minutes: $\frac{5}{60} = \frac{1}{12}$ of an hour

10 minutes: $\frac{10}{60} = \frac{2}{12} = \frac{1}{6}$ of an hour

15 minutes: $\frac{15}{60} = \frac{3}{12} = \frac{1}{4}$ of an hour

30 minutes: $\frac{30}{60} = \frac{6}{12} = \frac{1}{2}$ of an hour

45 minutes: $\frac{45}{60} = \frac{9}{12} = \frac{3}{4}$ of an hour

Minutes as a decimal of 1 hour

1 minute = $1 \div 60$ = $0.01\dot{6}$ of an hour

5 minutes = $5 \div 60$ = $0.083\dot{}$ of an hour

10 minutes = $10 \div 60$ = $0.1\dot{6}$ of an hour

15 minutes = 0.25 of an hour

30 minutes = 0.5 of an hour

45 minutes = 0.75 of an hour

(a) 20 minutes = $\frac{1}{3}$ hour; $20 \div 60 = 0.\dot{3}$

(b) $0.25 = \frac{1}{4}$

Worked example

(a) What is 1 hour 20 minutes as
  (i) a fraction
  (ii) a decimal of an hour?
  (i) $1\frac{20}{60} = 1\frac{1}{3}$ hour   (ii) $1\frac{1}{3} = 1.\dot{3}$ hour
(b) What is 4.25 hours in hours and minutes?
  0.25 hours = $60 \div 4 = 15$ minutes
  4.25 hours is 4 hours 15 minutes

Now try this

1 How many minutes are there in 5.4 hours?

2 Write 230 minutes as
  (a) a fraction
  (b) a decimal of an hour.

3 What is 5.75 hours in hours and minutes?

4 Which is longest: 35 hours, 2000 minutes or 1.5 days? Show your working.

Convert them all to the same unit, either minutes or hours.

Speed, distance, time

Speed is a measure of how fast something is travelling. It is the distance travelled per unit of time. Common units of speed are:

- km/h: kilometres per hour
- mph: miles per hour
- m/s: metres per second.

Speed is an example of a **rate of change**.

Formula triangle

This is the **formula triangle** for speed.

To find the formula you need, put your finger over the quantity that you want to work out.

Average speed = $\dfrac{\text{distance}}{\text{time}}$ $S = \dfrac{D}{T}$

Distance = speed × time $D = ST$

Time = $\dfrac{\text{distance}}{\text{speed}}$ $T = \dfrac{D}{S}$

Worked example

(a) A train completes a journey of 480 km in 4 hours. Calculate its average speed.

Average speed $= \dfrac{\text{distance}}{\text{time}}$

$= \dfrac{480}{4} = 120$ km/h

(b) A boat can travel at an average speed of 12 mph. How long will it take to complete a journey of 60 miles?

Time $= \dfrac{\text{distance}}{\text{speed}}$

$= \dfrac{60}{12} = 5$ hours

Check your units for speed. The distance is measured in km and time is measured in hours so the speed will be measured in km/h.

The speed is measured in miles per hour so the time will be in hours.

Minutes and hours

When calculating distance or time, you must make sure the **units match**.

Use this diagram to help when converting between minutes and hours.

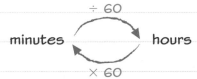

45 minutes = 45 ÷ 60 = 0.75 hours
2.4 hours = 2.4 × 60 = 144 minutes

For a reminder about converting units of time turn to page 54.

Worked example

A helicopter travels for 30 minutes at an average speed of 100 km/h. How far has it travelled?

Distance = average speed × time

$= 100 × 0.5$

$= 50$ km

Speed is given in km/h so the time must be converted to hours.

30 minutes $= \dfrac{30}{60}$ hours = 0.5 hours

Now try this

1 How long does it take a lion running at 20 m/s to cover 400 m?

2 Paul leaves his house at 08:00. He cycles 10 km to school at an average speed of 15 km/h. At what time does Paul arrive at school?

Work out the time it takes Paul, in hours, and then multiply by 60 to find the number of minutes.

Percentage change

Sales, reductions, depreciation, taxes, price increases and interest on savings are usually described in terms of percentage increase or decrease. To understand them you need to be able to find percentages of amounts.

Here are two methods for increasing or decreasing an amount by a given percentage.

 'Add or subtract' method

1. Work out the percentage of the amount.
2. Add or subtract this to or from the original amount.

£380 increased by 25%

25% of £380 = 380 ÷ 100 × 25 = £95
New amount: £380 + £95 = £475

£648 decreased by 5%

5% of £648 = 648 ÷ 100 × 5 = £32.40
New amount: £648 − £32.40 = £615.60

Turn to page 25 for a reminder about percentages of amounts.

 Decimal multiplier method

1. 100% is the original amount.
2. An **increase** of 30% = 100% + 30%
   = 130% = 1.3
3. A **decrease** of 15% = 100% − 15%
   = 85% = 0.85

Increase £565 by 30%

£565 × 1.3 = £734.50

Decrease £420 by 15%

£420 × 0.85 = £357

For a reminder on decimal–percentage equivalents turn to page 24.

Worked example

Use your preferred method to work these out.
(a) Decrease £1250 by 35%
1250 ÷ 100 × 35 = 437.50
1250 − 437.50 = £812.50
(b) Add 18% tax to £2400
2400 ÷ 100 × 18 = 432
2400 + 432 = £2832
(c) Save 25% on £90
90 ÷ 100 × 25 = 22.5
90 − 22.5 = £67.50

These show the 'add or subtract' method. Alternatively, you could use a decimal multiplier.

(a) 100% − 35% = 65% = 0.65
   1250 × 0.65 = £812.50
(b) Increase of 18% means 100% + 18%
   = 118% = 1.18
   2400 × 1.18 = £2832
(c) Decrease of 25% means 100% − 25%
   = 75% = 0.75
   90 × 0.75 = £67.50

These show the 'decimal multiplier' method. Alternatively, you could use the 'add or subtract' method.

(a) Work out 15% of £465 and add it to £465.
(b) Work out 2.5% of £1500 and add it to £1500.

Worked example

(a) A holiday costing £465 in June costs 15% more in July. What is the cost in July?
465 × 1.15 = £534.75
(b) A bank is offering 2.5% interest on savings. Rich has saved £1500. How much will he have after the interest is added?
1500 × 1.025 = £1537.50

Now try this

1 Ten years ago the population of a town was 85 500. It has increased by 8%. What is the current population?

2 Which offers better value for the same item: a 15% discount on a price of £250 or 18% tax added to a price of £180?

Ratios

Ratios compare quantities. You can **simplify ratios** and find **equivalent ratios** in the same way that you simplify fractions and find equivalent fractions.

	=	

3 : 1   6 : 2

Worked example

Group the equivalent ratios.
1:3   150:200   9:12   50:150   5:15   6:9
2:3   100:150   20:60   60:80   18:27   3:4
1:3 = 20:60 = 50:150 = 5:15
2:3 = 6:9 = 18:27 = 100:150
3:4 = 9:12 = 150:200 = 60:80

Simplify all the ratios using division and the HCF.
1:3, 2:3 and 3:4 are already simplified.

20:60   100:150   9:12
÷20 ⟍   ⟋ ÷20   ÷50 ⟍   ⟋ ÷50   ÷3 ⟍   ⟋ ÷3
1:3   2:3   3:4

For a reminder about factors turn to page 11 and for fractions turn to page 19.

Dividing in a given ratio

Follow these steps to divide an amount in a given ratio.

1. Add up all the ratio parts to find the total number of parts.

Share £100 between Abi and Isla in the ratio 2:3
2 + 3 = 5 parts

2. Divide the amount by the number of parts to find the value of **one** part.

£100 ÷ 5 = £20

3. Multiply each number in the ratio by the value of one part.

Abi: 2 × £20 = £40; Isla: 3 × £20 = £60

Worked example

A label says 80% cotton:20% acrylic. Write the ratio of cotton to acrylic in its simplest form.

cotton : acrylic
÷20 ⟍   80 : 20   ⟋ ÷20
        4 : 1

• Write the ratio in the order it is given.
• Simplify the ratio fully.

Worked example

A businessman gives away some of his wages to a food bank, to medical research and to a disaster relief agency in the ratio 1:2:3. Last year he gave away £2400. How much did each charity receive?

1 + 2 + 3 = 6 parts
£2400 ÷ 6 = £400
Food bank: 1 part = £400
Medical research: 2 parts = 2 × 400 = £800
Disaster relief: 3 parts = 3 × 400 = £1200
Check: £400 + £800 + £1200 = £2400 ✓

Problem solved!

1. Add to find the total number of parts.
2. Divide the amount by the number of parts to find the value of one part.
3. Multiply the value of one part by each of the numbers in the ratio.
4. Check your answer.

You'll need brilliant problem-solving skills to succeed in GCSE – get practising now! 💡

Oliver gets 3 shares. Use this to find how much one share is (90 ÷ 3 = ?). Multiply this amount by the total number of shares.

Now try this

1 Simplify the ratio 16:36

2 Toby, Isobel and Oliver share some money in the ratio 2:5:3. Oliver gets £90. How much money did they share?

Proportion

A proportion is a relationship between two numbers. You can write a proportion as a ratio, a percentage or a fraction.

| 2 | : | 3 | | 3 | : | 1 | | 5 | : | 3 |

$$\frac{2}{5} + \frac{3}{5} = 1 \qquad \frac{3}{4} + \frac{1}{4} = 1 \qquad \frac{3}{8} + \frac{5}{8} = 1$$

40% + 60% = 100%   75% + 25% = 100%   62.5% + 37.5% = 100%

For a reminder about ratios turn to page 57 and for equivalence turn to page 24.

Worked example

(a) The ratio of carrots to onions in a vegetable plot is 3:2. What fraction are carrots?

$$\frac{3}{3+2} = \frac{3}{5}$$

(b) In a class $\frac{1}{6}$ of students are left-handed. What is the ratio of left- to right-handed students in this class?

$1 - \frac{1}{6} = \frac{5}{6}$ are right-handed

left	:	right
$\frac{1}{6}$	:	$\frac{5}{6}$
1	:	5

(a) Add the parts of the ratio to find the total, to give the denominator of the fraction.

(b) Work out how many are right-handed. Then write the ratio in the order asked for.

Worked example

These ingredients make a pizza that serves 8 people:
  400 g bread dough
  120 ml tomato puree
  240 g grated cheese

(a) How much bread dough is needed for 12 people?
400 g × 1$\frac{1}{2}$ = 400 g + 200 g = 600 g

(b) Maisie has 120 g cheese. How many people can she make pizza for if she has enough of the other ingredients?

120 is half of 240

So halve the number of people: 8 ÷ 2 = 4 people

(c) Yasdi has 500 g dough, 160 ml tomato puree and 160 g cheese. Is that enough ingredients to serve 10 people?

400 ÷ 8 = 50 g dough per person
50 g × 10 = 500 g. Enough dough.
120 ÷ 8 = 15 ml tomato puree per person
10 × 15 = 150 ml. Enough tomato puree
240 ÷ 8 = 30 g cheese per person
30 g × 10 = 300 g. Not enough cheese.
No, he does not have enough cheese to make pizzas to serve 10 people.

(a) 12 is 1$\frac{1}{2}$ × 8 so multiply the recipe quantity of bread dough by 1$\frac{1}{2}$

(c) Work out how much he needs of each ingredient for one person and then multiply this by 10.

Now try this

1 60% of a group are female. What proportion of the group are male? Give your answer as a simplified fraction.

Convert the hours and minutes to minutes: 2.5 × 60 = 150 minutes and 1 hour 50 = 60 + 50 minutes. Divide the time by the number of components made to work out how long each worker takes to make 1 component.

2 Factory workers are making components. Which worker is the fastest, which two workers work at the same rate, and which worker is the slowest?

Worker	Number made	Time taken
A	15	2.5 h
B	20	3.5 h
C	18	3 h
D	12	1 h 50 min

Direct proportion

When two quantities are in direct proportion they both increase or decrease at the **same rate**. They are always in the same **ratio**.

Graphs of direct proportion always have the same shape: a straight line going through the origin (O, O).

The unitary method

You can sometimes solve problems by working out the cost of **one item**.

1 Divide to find the value of 1 item.

2 Multiply to work out cost of the new quantity.

3 theatre tickets cost £135.
How much will 8 tickets cost?

	tickets	cost	
÷3	3	£135	÷3
×8	1	£45	×8
	8	£360	

Worked example

A garden centre sells large and small plants.
(a) 15 large plants cost £75. How much will 11 large plants cost?

75 ÷ 15 = 5
11 × 5 = 55
11 large plants will cost £55

(b) Small plants are available in trays of 6 for £15 or boxes of 10 for £24.
Work out whether the box or the tray offers better value. Show all your working.

Tray   Box
15 ÷ 6 = 2.5   24 ÷ 10 = 2.40
£2.50 per plant   £2.40 per plant
The box is better value because £2.40 < £2.50

The number of large plants and the cost are in direct proportion. Work out how much 1 plant costs and then multiply that amount by 11. Make sure you give units (£) with your answer.

You need to show all your working in this question. Work out the cost of each plant in a tray of 6. Work out the cost of each plant in a box of 10. Write a short conclusion saying which one is better value and why.

(a) Multiply by the exchange rate:

        £1 $2.11
    × 560 (   ) × 560
        £560 $1181.60

(b) Changing back: divide by the exchange rate.

Worked example

Peyton is going on holiday to Australia. He finds this exchange rate online: £1 = $2.11
(a) He changes £560. How many Australian dollars does he receive?

560 × 2.11 = $1181.60

(b) Sally changes $560 into pounds using the same rate. How many pounds does she receive, to the nearest penny?

560 ÷ 2.11 = 265.4028436
            = £265.40

Now try this

1 Six identical notebooks cost £3.
  Work out
  (a) the cost of 12 of these notebooks
  (b) the cost of 1 of these notebooks
  (c) the cost of 7 of these notebooks.

Work out 120 ÷ 13.44 and then round to 2 d.p.

2 Emily is going on holiday to Sweden. She changes £450 into Swedish krona (kr) using the exchange rate £1 = 13.44kr
  (a) How many krona does she receive?
  (b) At the end of her holiday Emily changes 120kr back into pounds at the same exchange rate. How many pounds does Emily receive? Give your answer to the nearest penny.

Inverse proportion

Two quantities are in inverse proportion if one quantity **increases** at the same rate as the other quantity **decreases**.

Graphs of inverse proportion have a distinctive shape.

Average speed   Time taken

40 km/h   2 hours

×2   ÷2

80 km/h   1 hour

Speed and time taken to travel the same distance are in inverse proportion: if you go faster, you reach your destination more quickly!

Worked example

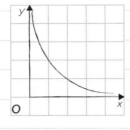

It takes 4 farmworkers 3 days to put up a fence around a field.

(a) How long would it take 6 workers to put up a similar fence?

$4 \times 3 = 12$ days' work in total

$12 \div 6 = 2$ days

(b) A group of workers took 4 days to put up a fence at the same rate. How many workers were there?

$12 \div 4$ days $= 3$ workers

More or less?

Inverse proportion problems often involve time, such as the number of people needed to complete a task in a given time. The number of people and the time are in inverse proportion:

- the more people, the less time
- the fewer people, the more time.

Problem solved!

First work out the number of days of work to do the task: 4 workers $\times$ 3 days $=$ 12 days

'Sense check' your answers: more workers will take less time.

You'll need brilliant problem-solving skills to succeed in GCSE – get practising now!

Constant product

If two quantities are in inverse proportion then their product will be **constant**.

$$\text{speed} = \frac{\text{distance}}{\text{time}} \qquad \text{so} \qquad d = s \times t$$

Speed (mph)	Time (h)	Speed × Time
10	6	60 miles
20	3	60 miles
30	2	60 miles
40	1.5	60 miles

The graph shows the inverse relationship between speed and time.

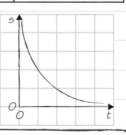

Now try this

1 3 cyclists travel 60 km. Copy and complete the table to show that speed and time are inversely proportional.

Speed (km/h)	Time (h)	Speed × Time
15	4	60 km
12		60 km
	2.5	60 km

2 It took 8 men a total of 6 hours to build a wall.

(a) How long would it take 3 men to build the same wall, working at the same rate?

(b) How long it would take 12 men to build the same wall, working at the same rate?

First work out how many hours it took 8 men to build the wall. $8 \times 6 = 48$ hours. Then divide this by the number of people each time.

Distance–time graphs

A distance–time graph shows how distance changes over time during a journey. The shape of the graph tells you about different parts of the journey. The slope shows the **speed** – the steeper the slope, the greater the speed.

Understanding distance–time graphs

The distance–time graph shows Tariq's journey. He drives from home to a friend's house where he stops for coffee, then he drives to a shopping centre where he shops for a while, and then he drives straight home.

The gradient of a section of a distance–time graph gives the speed at that time, because speed = $\dfrac{\text{distance}}{\text{time}}$.

For a reminder about gradient turn back to page 46, and for speed turn to page 55.

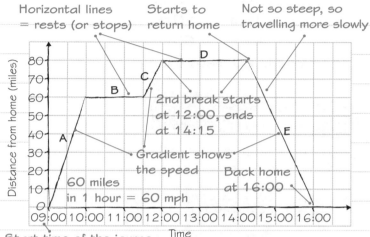

Horizontal lines = rests (or stops)   Starts to return home   Not so steep, so travelling more slowly

2nd break starts at 12:00, ends at 14:15

Gradient shows the speed

60 miles in 1 hour = 60 mph

Back home at 16:00

Start time of the journey   Time

Worked example

Look at the graph above.
(a) How many miles did Tariq travel?
60 + 20 + 80 = 160 miles
(b) At what time did Tariq begin his return journey?
14:15
(c) How long did Tariq spend shopping?
2 hours 15 minutes
(d) What was his speed during the third part of his journey?
20 miles ÷ $\frac{1}{2}$ hour = 40 miles ÷ 1 hour
= 40 mph
(e) In which section was Tariq driving fastest?
A: speed = 60 mph
E: speed = $\dfrac{80}{1.75}$ = 46 mph
He drove fastest in section A because 60 mph is faster than 40 mph and 46 mph.

Problem solved!

(a) Add the distances travelled in sections A, C and E. Don't forget the return journey!
(b) Section E starts halfway between 14:00 and 14:30.
(c) Section D starts at 12:00 and finishes at 14:15.
(e) Work out the speeds in the other sections when he was driving. Section E takes $1\frac{3}{4}$ hours = 1.75. Compare the speeds.

You'll need brilliant problem-solving skills to succeed in GCSE – get practising now!

Now try this

The graph shows Lara's journey to the cinema. She travels by bus to town, waits to meet her friend, then they walk to the cinema together.
(a) How long did Lara's bus journey to town take?
(b) How long was the film?
After the film they walk to the bus stop. Lara waits for 10 minutes before catching the bus home, travelling at the same speed as her journey into town.

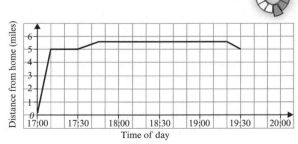

(c) Copy and complete the graph to show Lara's journey back home.

Maps and scales

You can use the scales on maps and scale drawings to work out distances in real life.

This is a scale drawing of a plane. You can use the scale to find the length of the real plane. The scale drawing is 6 cm long so in real life the length of the plane is

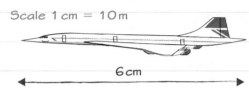

Scale 1 cm = 10 m

```
        drawing   real life
       ╭ 1 cm      10 m ╮
  × 6  ╰ 6 cm      60 m ╯  × 6
```

6 cm

The plane is 60 metres long.

Worked example

The diagram shows a scale drawing of a village.
Scale: 1 cm = 3 km

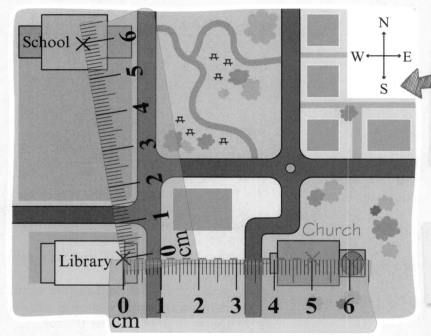

1. Use a ruler to measure the distance on the map from the library to the school. Distance on map = 6 cm

2. Now use the scale to work out the real distance.

```
        map      real
       ╭ 1 cm    3 km ╮
  × 6  ╰ 6 cm    18 km ╯  × 6
```

```
        map      real
       ╭ 1 cm    3 km ╮
  × 5  ╰ 5 cm    15 km ╯  × 5
```

(a) What is the distance between the school and the library? 3 × 6 = 18 km

(b) The church is 15 km east of the library. Mark its position with a cross.

Now try this

The map shows the attractions at a zoo.

(a) Work out the real distance between the coffee shop and the lion enclosure.

(b) Paula says the distance between the giraffe enclosure and the lion enclosure is less than 250 m. Use the map to show whether she is correct.

(c) The distance between the giraffe enclosure and the children's area is 400 m. What is the distance on the map?

Scale: 1 cm = 50 m

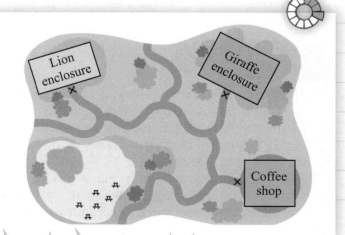

Proportion problem-solving

You can use proportion to solve problems which involve changing one unit of measurement into a different unit and when calculating rates of change.

To revise proportion look at pages 58 and 59. For metric measures look at page 53 and to revise time and speed look at pages 54 and 55.

Worked example

Which is the faster speed, 36 m/s or 154 km/h? You must show your working.

36 m/s = 36 ÷ 1000 × 60 × 60 = 129.6 km/h

154 km/h is faster than 36 m/s

To compare the speeds, the units must be the same. Either convert 36 m/s to km/h, or convert 154 km/h to m/s:

154 × 1000 ÷ 60 ÷ 60 = 42.8 m/s

In part (b) rearrange the formula to make m the subject. In part (c), convert 2691 kg into grams (× 1000) and convert m³ into cm³ (× 1 000 000).

Worked example

(a) A gold bracelet has a mass of 57.9 g and a volume of 3 cm³. Work out the density of the gold using the formula

$$\text{density} = \frac{\text{mass}}{\text{volume}}$$

$d = \dfrac{57.9}{3} = 19.3 \text{ g/cm}^3$

(b) A silver necklace has a density of 10.5 g/cm³ and a volume of 1.4 cm³. What is its mass?

$m = dv = 10.5 \times 1.4 = 14.7 \text{ g}$

(c) The density of solid granite is 2691 kg/m³. Convert this density into g/cm³.

$d = \dfrac{2691 \times 1000}{1 \times 1\,000\,000}$

$= 2691 \div 1000 = 2.691 \text{ g/cm}^3$

Worked example

Garage A sells petrol for £1.09 per litre. Garage B sells petrol for £4.99 per gallon.
Which garage sells the cheaper petrol?
You must show your working.
Use 1 gallon = 4.55 litres

£1.09 × 4.55 = 4.9595 = £4.96 per gallon

Petrol is cheaper at garage A because £1.09 per litre converts to £4.96 per gallon and this is cheaper than £4.99 per gallon at garage B.

Problem solved!

You **must** show how you worked out the answer. To compare prices you need to have the quantities in the same units. This solution compares the cost per gallon but you could have compared the cost per litre by dividing the cost of petrol at garage B (£4.99 ÷ 4.55 = £1.10 per litre).
Make sure you write a statement to answer the question.

You'll need brilliant problem-solving skills to succeed in GCSE – get practising now!

Now try this

1 Which metal is more dense: a 20 000 kg iron cube of volume 2.5 m³ or a 36 g cube of copper of volume 4 cm³?

2 Mark travelled 450 miles in 7.25 hours. Dee travelled the same distance at 100 km/h. If they set off at the same time, who arrived first? Show how you know. Use the approximate conversion 1 mile ≈ 1.6 km.

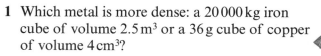

Work out the density of both metals. Then convert one of them so that both densities are in the same units, either g/cm³ or kg/m³. Take care when multiplying or dividing by 1000.

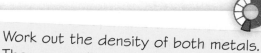

Work out Mark's speed in mph and change 100 km/h into mph – who is faster?

3D shapes

You need to recognise and learn the names of certain 3D shapes.

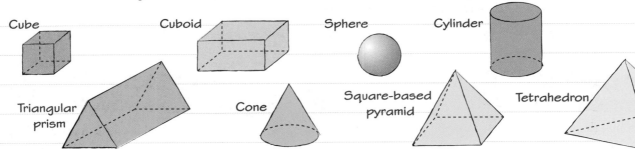

Cube   Cuboid   Sphere   Cylinder

Triangular prism   Cone   Square-based pyramid   Tetrahedron

Drawing 3D shapes

You can use isometric paper to draw 3D shapes in 2D.

This 3D shape has been drawn on isometric paper.

This cuboid is 2 units wide, 3 units long and 2 units high.

Prisms

A prism is a 3D shape that has the same cross-section throughout its length.

The cross-section is always perpendicular (at right angles) to its length.

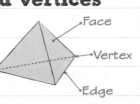

Worked example

(a) Write the name of this shape.
Triangular prism
(b) How many edges does the shape have?
9 edges

Faces, edges and vertices

This tetrahedron has 4 faces, 6 edges and 4 vertices.
The plural of vertex is vertices.

Face
Vertex
Edge

The net of a 3D solid is a 2D shape that folds to make up the solid.

Worked example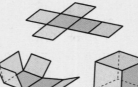

Which of these is the net of a cube?

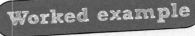

A   B   C

B

Now try this

1 (a) Write the names of these shapes.
   (b) Which shape is **not** a prism?
   (c) Which shape has 8 vertices?
   (d) What shape is the cross-section of shape (iii)?

(i) 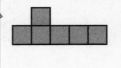   (ii)   (iii)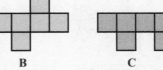

2 Which of these is a net of a tetrahedron?

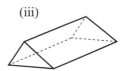

How many faces does a tetrahedron have? What shape are they?

A   B   C

Volume

The **volume** of a 3D shape is the amount of space it takes up. Volume is measured in cubed units such as cubic millimetres (mm^3), cubic centimetres (cm^3) or cubic metres (m^3).

Counting cubes

You can find the volume of a 3D shape made from cubes by counting the cubes in the shape.

Volume = $1 cm^3$ Volume = $6 cm^3$

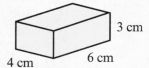

This shape is made from six $1 cm^3$ cubes.

Worked example

Work out the volume of this shape.

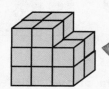

Remember to count the hidden cubes.

There are 8 cubes in the front section.
There are 2 sections.
Altogether there are 2 × 8 cubes.
The volume is $16 cm^3$

Worked example

Work out the volume of this cuboid.

3 cm

4 cm   6 cm

Volume   = length × width × height
      = 6 × 4 × 3
      = $72 cm^3$

Volume of a cuboid

You need to learn the formula for the volume of a cuboid.

height
width
length

Volume = length × width × height

Worked example

This shape is made from cuboids. Find its volume.

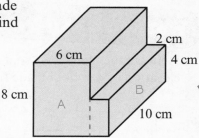

6 cm
2 cm
4 cm
8 cm
A   B
10 cm

Volume A = 8 × 6 × 10 = $480 cm^3$
Volume B = 4 × 2 × 10 = $80 cm^3$
Total volume = 480 + 80 = $560 cm^3$

Split the shape into two cuboids.

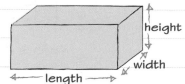
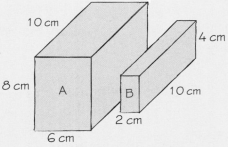

10 cm
4 cm
8 cm
A
B   10 cm
2 cm
6 cm

Calculate the volume of each cuboid. Then add them to get the total volume. Write the units in your answer.

Now try this

1 This shape is made of $1 cm^3$ cubes. What is its volume?

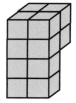

Always write the units in your answer.

2 This prism is made from cuboids. Find its volume.

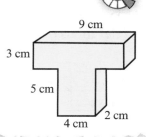

9 cm
3 cm
5 cm
4 cm
2 cm

Split the prism into cuboids.

Plans and elevations

When you look at a 3D shape from different directions you see 2D shapes. These are called plans and elevations. For a reminder about 3D shapes turn to page 64.

plan

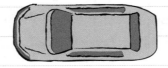
plan

front elevation

side elevation

The plan is the view from above. The side elevation is the view from the side. The front elevation is the view from the front.

Worked example

The diagram shows a solid shape.
On the grid, draw a plan view and front and side elevations of the shape.

Plan

Front elevation   Side elevation

Shade what you will see from each view first. This will help you imagine each 2D shape.

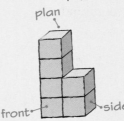

plan

plan

side elevation

front   side

Draw lines within the plan view and side elevation to show where there is a change of height or depth.

Imagine the plan view. From above you will see 6 cubes. From the side you will see 3 cubes. From the front you will see 2 cubes. If you need more help, make the shape with cubes to check.

Worked example

Here are the plan view, front elevation and side elevation of a cuboid made from cubes.

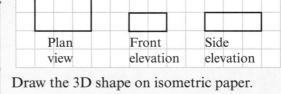

Plan view   Front elevation   Side elevation

Draw the 3D shape on isometric paper.

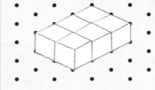

Now try this

Draw the plan view, front and side elevations of this 3D shape.

Measuring and drawing angles

You use a protractor to draw and measure angles in degrees. Knowing the names and approximate sizes of the different types of angle will help you to measure and draw angles accurately.

acute < 90   right 90°   obtuse between 90° and 180°   reflex between 180° and 360°

1 Measuring angles

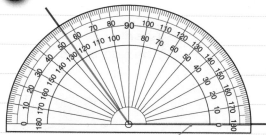

Use the scale that starts from 0 along the base line of the angle. Here, use the inside scale.

Place the centre of the protractor at the point where the two lines meet.

Line up the zero line with one line of the angle.

Read the size of the angle off the scale.

This angle is 125°.

Worked example

(a) Estimate and name the size of each angle by comparing them with the diagrams at the top of the page.
(b) Check the size of each angle by measuring.

110°   235°   45°   reflex   obtuse   acute

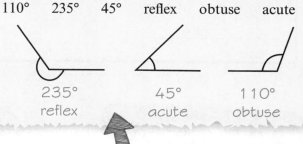

235° reflex   45° acute   110° obtuse

Estimate the size of the angle first. This helps you check your answer is sensible.

2 Drawing angle

Draw an angle of 18°.

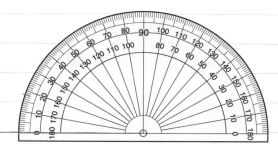

Use a ruler to draw a line.

Place the centre of the protractor at one end of the line.

Find 18° on the scale that starts at zero on your line. This time, use the outside scale. Draw a dot or small mark at 18°.

Using a ruler, join the end of the line to the mark.

Draw the angle arc and label your angle 18°.

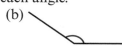

18°

Now try this

1 Draw an angle of 135°.
2 Measure and name each angle.
  (a)   (b)

3 Is this angle 65° or 115°? Explain how you know.

Angles 1

Angles are named using the three letters of the lines that make the angle.

The point where the angle is formed is **always** the middle letter.

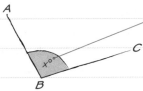

The angle $x°$ can also be written as ABC

Angle facts

You can use these angle facts to work out missing angles.

1 Angles on a straight line add up to 180°.
$a + b = 180°$

2 Angles at a point add up to 360°.
$c + d + e = 360°$

3 Vertically opposite angles are equal.
$f = g$ $h = i$

Worked example

Work out the sizes of the angles marked with letters. Give reasons for your answers.

(a)   (b)

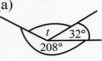

(a) $t + 32° + 208° = 360°$
$t = 360 - 240$
$t = 120°$
Angles at a point add up to 360°.

(b) $u + 38° = 180°$
$u = 180 - 38 = 142°$
Angles on a straight line add up to 180°.
$w = 142°$ Vertically opposite angles are equal.
$v = 38°$ Vertically opposite angles are equal.

Worked example

Work out the size of the larger angle.

$5x + x = 180°$
$6x = 180°$
$x = 30°$ so $5x = 150°$

Problem solved!

1. Use an angle fact: angles on a straight line add up to 180°.
2. Set up and solve an equation.
3. Substitute your value for x into the larger angle.

You'll need brilliant problem-solving skills to succeed in GCSE – get practising now!

Now try this

Find the size of the smallest angle.

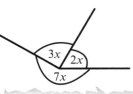

Use the angle fact about angles at a point.

Angles 2

Learn these angle facts.

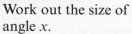
$a + b + c = 180°$
Angles in a triangle add up to 180°.

$d + e + f + g = 360°$
Angles in a quadrilateral add up to 360°.

Worked example

Work out the size of angle x.

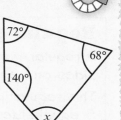

$x = 360 - 280 = 80$
$x = 80°$
Angles in a quadrilateral add up to 360°.

Worked example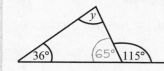

Work out the size of angle y.

Write any angles you have worked out on the diagram.

$180 - 115 = 65$
Angles on a straight line add up to 180°.
$y = 180 - (65 + 36) = 79$
$y = 79°$
Angles in a triangle add up to 180°.

Angles on parallel lines

Parallel lines are always the same distance apart. They never meet. They are marked with arrows.

Learn these angle facts about parallel lines.

 ①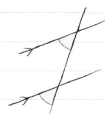

②

③

$a + b = 180°$

Corresponding angles are equal.

Alternate angles are equal.

Co-interior or **allied** angles add up to 180°.

Worked example

Work out the sizes of the angles marked with letters. Give reasons for your answers.

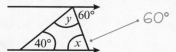

$x = 60°$ Alternate angles are equal
$y = 180 - (40 + 60) = 80$
$y = 80°$ Angles in a triangle add up to 180°

Work through the problem one step at a time, writing your answers and reasons clearly.

Now try this

Work out the sizes of the angles marked with letters. Give reasons for your answers.

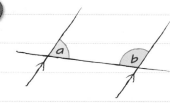

Use the angle facts about angles on parallel lines.

 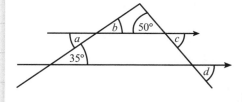

Angles in polygons

Polygons are 2D shapes with straight sides. You need to be able to identify interior and exterior angles in polygons.

Exterior angle + interior angle = 180°

Angles on a straight line add up to 180°

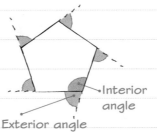

Interior angle

Exterior angle

Formulae for angles in polygons

Learn these formulae for a polygon with n sides.

1 Sum of interior angles $= (n - 2) \times 180°$

2 Sum of exterior angles $= 360°$

Worked example

(a) What is the sum of the interior angles of a heptagon?

A heptagon has 7 sides so $n = 7$

Sum of interior angles $= (n - 2) \times 180°$
$= (7 - 2) \times 180°$
$= 5 \times 180° = 900°$

(b) The sum of the interior angles of a polygon is 1080°. How many sides does it have?

$(n - 2) \times 180° = 1080°$
$n - 2 = 1080 \div 180$
$n - 2 = 6$
$n = 8$

The shape has 8 sides.

For a reminder on solving equations turn to pages 39 and 40.

Regular polygons

A regular polygon has equal sides and equal angles.

This regular pentagon has 5 equal sides and 5 equal exterior angles.

$\frac{360°}{5}$

72°

$n = 5$

The sum of the exterior angles of any polygon is 360°.

So each exterior angle of a regular polygon of n sides is $\dfrac{360°}{n}$

Each exterior angle of this pentagon is $\dfrac{360°}{n} = 72°$

Each interior angle is $180° - 72° = 108°$

Names of polygons

Learn the names of these polygons:

Quadrilateral – 4 sides

Pentagon – 5 sides

Hexagon – 6 sides

Heptagon – 7 sides

Octagon – 8 sides

Nonagon – 9 sides

Decagon – 10 sides

Worked example

(a) Work out the exterior angle of a regular decagon.

Sum of exterior angles is 360°.
So exterior angle $= 360° \div 10 = 36°$

(b) Calculate one of the interior angles.

Exterior angle + interior angle = 180°
Interior angle $= 180 - 36 = 144$
Interior angle is 144°

Now try this

1 Find the sum of the interior angles of a 12-sided shape.

Use the formula
Sum $= (n - 2) \times 180°$

2 (a) Calculate one of the exterior angles in a regular nonagon.

(b) Calculate one of the interior angles.

Interior angle + exterior angle = 180°

Pythagoras' theorem

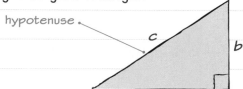

Pythagoras' theorem is a rule that you can use to find the length of a missing side in a right-angled triangle.

hypotenuse

c

b

a

$a^2 + b^2 = c^2$

In a right-angled triangle, the longest side is called the **hypotenuse**. It is always opposite the right angle.

Pythagoras checklist
- ✓ short^2 + short^2 = long^2
- ✓ Right-angled triangle.
- ✓ Lengths of two sides known.
- ✓ Length of third side missing.

Worked example

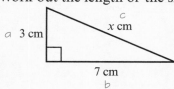

Work out the length of the side labelled x.

c

x cm

a 3 cm

7 cm
b

$c^2 = a^2 + b^2$
$x^2 = 3^2 + 7^2$
   $= 9 + 49 = 58$
$x = \sqrt{58} = 7.615\,77$
$x = 7.6$ cm (1 d.p.)

1. Label the longest side of the triangle c.
2. Label the other two sides a and b. (It doesn't matter which is a and which is b.)
3. Substitute your values into the formula $c^2 = a^2 + b^2$
4. Solve the equation, remembering to take the square root at the end.
5. Include the units in your final answer.

Problem solved!

1. Draw a diagram.
2. Label the longest side c and label the other two sides a and b.
3. Substitute the values into the formula $a^2 = c^2 - b^2$
4. Rearrange and solve the equation.

> You'll need brilliant problem-solving skills to succeed in GCSE – get practising now! 💡

Worked example

A ladder 5 m long is placed next to a wall. The base of the ladder is 2 m from the bottom of the wall.
How high up the wall does the ladder reach?

$a^2 = c^2 - b^2$
$a^2 = 25 - 4 = 21$
$a = \sqrt{21} = 4.58257$
$a = 4.6$ m (1 d.p.)

5 m
c

a

2 m
b

Now try this

Work out the length of the diagonal in this square.

> Draw in the diagonal. Is this a shorter side or the hypotenuse?

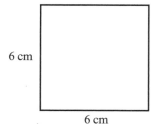

6 cm

6 cm

Drawing triangles

You will use different tools for drawing triangles. If you are asked to **draw** a triangle with angles of given sizes, you will need a protractor and a ruler. If you are asked to **construct** a triangle you will need a pair of compasses and a ruler.

Worked example

Use a ruler and protractor to make an accurate drawing of this triangle.

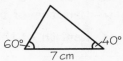

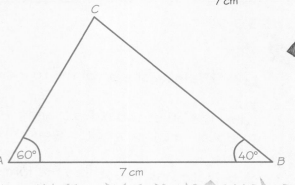

1. Use a ruler to draw a line of length 7 cm first.
2. Use your protractor to draw an angle of 60° at A.
3. Then draw an angle of 40° at B.
4. Point C is where these lines cross.

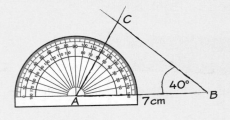

---

1. Use a ruler to draw and label a line of length 6 cm. Label the ends A and B.
2. Open your compasses to 4 cm. Place the point at A and draw an arc.
3. Open your compasses to 5 cm. Place the point at B and draw an arc.
4. Label the point where the two arcs cross as C.

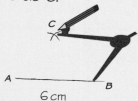

5. Join up the vertices of your triangle.

Worked example

Use compasses and a ruler to construct a triangle with sides of length 4 cm, 5 cm and 6 cm.

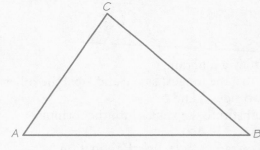

Now try this

1 Use a ruler and protractor to make an accurate drawing of this triangle.

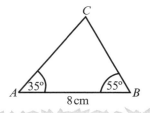

2 Use compasses and a ruler to construct this isosceles triangle.

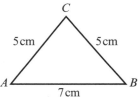

Constructing perpendicular lines

You need to know how to use compasses and a ruler to **construct** the **perpendicular bisector** of a line.

Perpendicular means 'at right angles'. Bisect means 'cut in half'.

A perpendicular bisector is the line that cuts another line in half at right angles.

Worked example

Use a pair of compasses and a ruler to construct the perpendicular bisector of a line 6 cm long.

   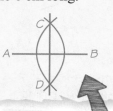

Draw a line of length 6 cm and label it AB.

Open your compasses to a radius that is greater than half the length of AB. Place your compass point on A, and draw an arc above and below the line AB.

With the compasses at the same radius, put the point on B and draw a second arc that cuts the first arc at C and D.

Use a ruler to join C and D. The line CD is the **perpendicular bisector** of the line AB.

Worked example

The diagram shows a scale drawing of a garden.
A gardener wants to plant a tree at the same distance from A as from C.
Draw the perpendicular bisector of the line between points A and C to show where the tree can be planted.

Draw a line between A and C.

Now construct the perpendicular bisector.

Everything in red is part of the answer.

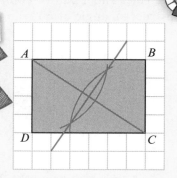

Now try this

1 (a) Draw a line AB 10 cm long.

   (b) Using a ruler and compasses, construct the perpendicular bisector of AB.

   (c) Use a ruler and protractor to check that it bisects your line at a right angle.

   (d) Mark a point × on your perpendicular bisector. Check that it is the same distance from A and B.

2 The diagram shows a scale drawing of a car park. A path is an equal distance from corners Q and S. Draw the perpendicular bisector of the line between points Q and S to show the path.

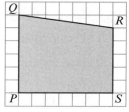

Constructing angles

You can construct angles of 90°, 45° and 60° using only a ruler and a pair of compasses. You can do this by bisecting angles. Remember, to bisect means to cut exactly in half.

For a reminder about bisectors turn to page 73.

Worked example

Show all your construction lines and marks. Leave arcs in place.

Use compasses and a ruler to bisect this angle of 50°.

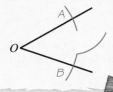

Open your compasses and place the point at O. Draw an arc that cuts both arms of the angle.

Reduce the radius of your compasses slightly. Place the compass point at A and draw another arc inside the angle.

Now keep your compasses set to the same radius. Place the compass point at B and draw another arc inside the angle. It should cross the first arc. Label this point X.

Use a ruler to join O and X. The line OX bisects the angle AOB.

Worked example

Use a ruler and a pair of compasses to construct an angle of 60°.

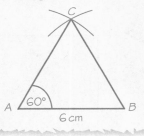

This equilateral triangle has sides of length 6 cm and interior angles of 60°.

If you need to construct an angle of 60° simply construct an equilateral triangle because all the angles in an equilateral triangle are 60°.

To revise constructing triangles look at page 72.

Constructing angles

✓ To construct an angle of 90°, draw a straight line and construct the perpendicular bisector.

✓ To construct an angle of 45°, construct an angle of 90° and then bisect the 90° angle.

Now try this

1 Use a ruler and a pair of compasses to bisect this angle of 70°.

2 Use a ruler and a pair of compasses to construct an angle of 45°.

Remember to leave in your arcs to show that you have used a pair of compasses.

Perimeter and area

Perimeter

Perimeter is the distance around the edge of a 2D shape. To work out the perimeter of a shape, you add up the lengths of all the sides.

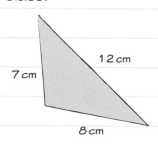

Perimeter = 8 cm + 7 cm + 12 cm
= 27 cm

Work out the perimeter of this rectangle.

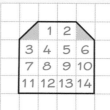

6 + 8 + 6 + 8 = 28
Perimeter = 28 cm

Work out the missing lengths first.

The opposite sides of a rectangle are equal so you can write these lengths on the diagram.

Area

Area is the amount of space inside a 2D shape.

Area is measured in squared units.

The area of one square is 1 cm². You say 'one square centimetre' or 'one centimetre squared'

Area = 14 cm²

You can work out the area of a shape drawn on centimetre squared paper by counting squares.

Work out the area of this shape.

Area = $14 + \frac{1}{2} + \frac{1}{2}$
= 15 cm²

Everything in red is part of the answer.

Count the whole squares and the half squares. Two half squares make one whole square.

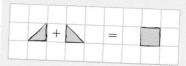

Golden rule

Always write the units with your answer.

- Units of perimeter include mm, cm, m and km.

- Units of area include mm², cm², m² and km².

1 This shape is drawn on a grid of centimetre squares.

(a) Find the perimeter of the shape.

(b) Find the area of the shape.

2 (a) Draw a shape with an area of 20 cm² on centimetre-squared paper.

(b) Find the perimeter of your shape.

3 The length of each side of a regular pentagon is 6 cm. Work out the perimeter of the pentagon.

All the sides of a regular polygon are the same length.

Area of rectangles and triangles

You need to know the formulae for the area of a rectangle and the area of a triangle.

1 Area of a rectangle

Area = length × width

$A = lw$

2 Area of a triangle

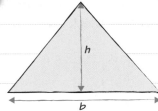

The vertical height, h, is perpendicular to the base of the triangle, b.

Area = $\frac{1}{2}$ × base × height

$A = \frac{1}{2}bh$

Worked example

The diagram shows the floor of a rectangular room.
(a) Work out the area of the room.

Area = $l \times w$
     = $5 \times 4 = 20\,m^2$

4 m

5 m

(b) Carpet costs £12 per m². How much would it cost to buy carpet for the room?

$20 \times 12 = 240$
Cost of the carpet is £240

Units

✓ Check the lengths are all in the same units.

✓ Remember to give units in your answer.

✓ Lengths in cm will give area units in cm².

✓ Lengths in m will give area units in m².

$\frac{1}{2} \times 15 \times 6$ is the same as $15 \times 6 \div 2$

Worked example

Work out the area of this triangle.

6 cm

15 cm

Area = $\frac{1}{2}bh$

     = $\frac{1}{2} \times 15 \times 6 = \frac{1}{2} \times 90$
     = $45\,cm^2$

Now try this

1 On centimetre-squared paper, draw two different rectangles, each with an area of 12 cm².

2 A square has a perimeter of 16 cm. Work out the area of the square.

3 The diagram shows a triangle of height 40 mm.

   (a) What is the height in cm?

   (b) Work out the area of the triangle.

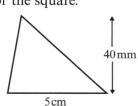

40 mm

5 cm

Area of parallelograms and trapeziums

You can use formulae to calculate the areas of parallelograms and trapeziums.

 Parallelogram

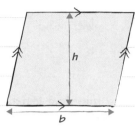

The vertical height, h, is at right angles to the base, b.

Area = base × height
$$A = bh$$

 Trapezium

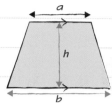

The vertical height, h, is the perpendicular distance between the parallel sides, a and b.

Area = $\frac{1}{2}$ × (sum of parallel sides) × height
$$A = \frac{1}{2}(a + b)h$$

Worked example

(a) Work out the area of this parallelogram.

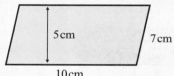

$A = bh$
$\quad = 10 \times 5$
$\quad = 50\,cm^2$

(b) Work out the perimeter of the parallelogram.

Perimeter = 10 + 10 + 7 + 7
$\qquad\qquad = 34\,cm$

(a) The vertical height is 5 cm.

Worked example

A garden is in the shape of a trapezium. Work out the area of the garden.

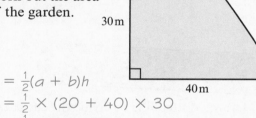

$A = \frac{1}{2}(a + b)h$
$\quad = \frac{1}{2} \times (20 + 40) \times 30$
$\quad = \frac{1}{2} \times 60 \times 30$
$\quad = 900\,m^2$

The trapezium has a right angle so the vertical height is 30 m.

Remember to work out the brackets first.

Now try this

1 The diagram shows a trapezium with a height of 6 cm.

Find the area of the trapezium.

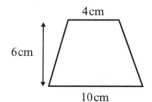

2 This parallelogram has an area of 40 cm².

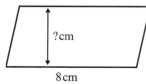

What is its height?

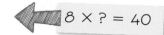 8 × ? = 40

Compound shapes

You can work out the area and perimeter of complex shapes by splitting them into simpler shapes.

Area = rectangle + triangle

Area = rectangle − triangle

Area = large rectangle − small rectangle

Worked example

Find the area of this shape.

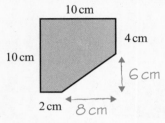

Area of square = 10 × 10 = 100 cm²
Area of triangle = 6 × 8 ÷ 2 = 24 cm²
Area of shape = 100 − 24 = 76 cm²

Golden rule

Always work out any missing lengths before calculating the area or perimeter.

You need to find the missing lengths and write them on the diagram.
10 − 4 = 6 cm   10 − 2 = 8 cm

Problem solved!

1. Draw a dotted line to divide the car park into two rectangles.
2. Label your two rectangles *A* and *B*.
3. Use the information in the question to find out the missing lengths.
   40 m − 30 m = 10 m   50 m − 10 m = 40 m
   Write these lengths on your diagram.
4. Find the area of each shape, then add them.

You'll need brilliant problem-solving skills to succeed in GCSE – get practising now!

Worked example

The diagram shows a car park.

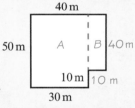

(a) Work out the area of the car park.
Area of A = 50 × 30 = 1500 m²
Area of B = 10 × 40 = 400 m²
Total area = 1500 + 400 = 1900 m²
(b) Calculate the perimeter of the car park.
50 + 40 + 40 + 10 + 10 + 30 = 180
Perimeter = 180 m

Now try this

(a) Find the missing height *h*.
(b) Work out the area of the shape.

What shapes can you split it into?

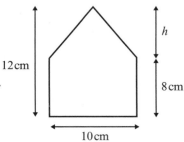

Circumference

You need to know these definitions of parts of a circle.

Circumference is the perimeter of a circle.

Diameter is the distance across the circle through the centre.

Radius is the distance from the centre to any point on the circumference. It is half of the diameter.

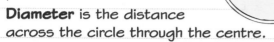

Radius = $\frac{1}{2}$ × diameter　　$r = \frac{d}{2}$

Diameter = 2 × radius　　$d = 2r$

They are all distances, so are measured in units such as mm, cm, m and km.

Here are two formulae you can use to calculate the circumference.

① Circumference = π × diameter
　　　　$C = \pi d$

② Circumference = 2 × π × radius
　　　　$C = 2\pi r$

π

π is the Greek letter 'pi'.

π = 3.1415926...

You can round π to 3.142 in calculations.

A scientific calculator will have a button for π. You might need to press the SHIFT key first.

If your calculator leaves π in the answer, press the [S↔D] button to get your answer as a decimal.

Golden rule

Write whether you have the radius or the diameter first so that you use the correct formula.

You may be asked to write your answer in terms of π. Just calculate with the numbers and leave π in your answer.
$C = 2 \times \pi \times 5 = 10\pi$

Worked example

Work out the circumference of this circle.
Use π = 3.142.
Give your answer to 1 decimal place.

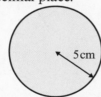
5cm

$r = 5\,cm$
Circumference = $2\pi r$
　　　　　　 = $2 \times 3.142 \times 5$
　　　　　　 = 31.42
　　　　　　 = $31.4\,cm$ (1 d.p.)

Worked example

Calculate the diameter of a circle with circumference 15 m. Write your answer to 1 decimal place.
$C = \pi d$
$15 = \pi d$
$d = 15 \div \pi = 4.7746... = 4.8\,m$ (1 d.p.)

You need to rearrange the equation $C = \pi d$ to make d the subject.
$C = \pi d\ (\div\ \pi)$
$d = C \div \pi$

For a reminder about rearranging formulae turn to page 43.

You need to find the radius – use the correct formula.

Now try this

1 Work out the circumference of a circle with
　(a) radius 3 cm
　(b) diameter 7 cm.

2 A circle has a circumference of 20 cm. Work out the radius of the circle.

Area of circles

You need to learn the formula for the area of a circle.

Area = $\pi \times$ radius^2

$A = \pi \times r \times r = \pi r^2$

You always use the **radius** when finding the area of a circle.

If you are given the diameter, divide by 2 to get the radius.

Worked example

Work out the area of this circle. Give your answer to 1 decimal place.

9 cm

$r = 9\,cm$

Area $= \pi r^2$

    $= \pi \times 9^2 = \pi \times 81$

    $= 254.4690...$

    $= 254.5\,cm^2$ (1 d.p.)

> Remember that squaring is multiplying the number by itself.
> 9^2 means 9×9.

> Write at least 5 digits from your calculator display before rounding.
> Make sure you include the units in your answer. The radius is given in cm so the area will be cm^2.

Worked example

Work out the area of this circle. Leave your answer in terms of π.

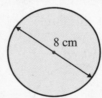

8 cm

$d = 8\,cm$

radius $= 8 \div 2 = 4\,cm$

Area $= \pi r^2$

    $= \pi \times 4^2 = \pi \times 16$

    $= 16\,\pi\,cm^2$

> To find the area of a circle you need the radius.
> The radius is **half** the diameter.

> Write the number in front of π when leaving the answer in terms of π .

Now try this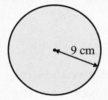

1 Work out the area of each circle.

  (a)

20 cm

  (b)

5 m

> Shaded region = area of − area of
>   square   circle

2 The diagram shows a circle and a square. The circle touches the edges of the square. Work out the area of the shaded region.

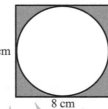

8 cm

8 cm

Circles problem-solving

You may be asked to solve problems that involve finding the area and perimeter of parts of circles.

For a reminder about circumference and area of circles turn to pages 79 and 80.

Worked example

Work out the area of the quarter circle.
Use $\pi = 3.142$.
Give your answer to 1 decimal place.

12 cm

$r = 12\,cm$
Area of circle $= \pi r^2$
$\qquad\qquad\quad = 3.142 \times 12^2 = 3.142 \times 144$
$\qquad\qquad\quad = 452.448\,cm^2$
Area of quarter circle $= 452.448 \div 4$
$\qquad\qquad\qquad\qquad\quad = 113.112$
$\qquad\qquad\qquad\qquad\quad = 113.1\,cm^2$ (1 d.p.)

Golden rules
✓ Always show your working.
✓ Write the formula before substituting in any numbers.

Problem solved!

1. Find the area of the whole circle.
2. Divide the area by 4 to get the area of the quarter circle.
3. Write the units in your answer.

You'll need brilliant problem-solving skills to succeed in GCSE – get practising now!

Problem solved!

1. Write whether you have the radius or the diameter so you can decide which formula to use.
2. Find the circumference of the whole circle.
3. Divide the circumference by 2 to get the length of the curved section of the semicircle.
4. Perimeter is the total distance around the outside of a shape.
   Perimeter = curved section + diameter
5. Don't forget to round your answer at the end of the question.

You'll need brilliant problem-solving skills to succeed in GCSE – get practising now!

Worked example

This flower bed is in the shape of a semicircle. Work out the perimeter of the semicircle.
Use $\pi = 3.142$.
Give your answer to 1 decimal place.

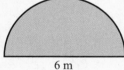

6 m

$d = 6\,m$
Circumference of circle $= \pi d$
$= 3.142 \times 6 = 18.852\,m$
Curved section of perimeter
$= 18.852 \div 2 = 9.426\,m$
Total perimeter of semicircle $= 6 + 9.426$
$= 15.426 = 15.4\,m$ (1 d.p.)

Now try this

This semicircle has a diameter of 12 cm.
(a) Work out the perimeter of the semicircle.
(b) Work out the area of the semicircle.

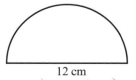

12 cm

Which formula are you going to use? Do you need the radius or the diameter?

Reflection

A **reflection** is a type of **transformation**. A shape can be reflected in a **mirror line**. The shape before the reflection is called the **object**, the shape after the reflection is called the **image**.

Shape A has been reflected in the mirror line to give shape B.

Reflected shapes are **congruent**.

To revise congruent shapes turn to page 86.

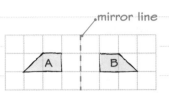

Doing reflections

You can use tracing paper to reflect shapes and check your answers.

Mark a cross on the mirror line.

Trace the object including the mirror line and cross.

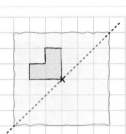

Turn the tracing paper over, lining up the mirror lines and the cross.

Trace the shape in the new position.

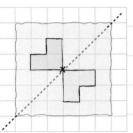

Worked example

Reflect shape A in the mirror line.

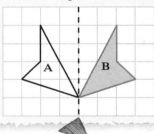

> Everything in red is part of the answer.

⬆ The mirror line is the line of symmetry. A dotted line is used to represent the mirror line.

Worked example

Describe fully the single transformation that maps shape A onto shape B.

Reflection in y-axis

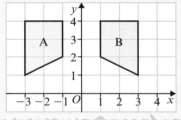

⬆ To 'fully' describe this you need to state that it is a reflection and describe the mirror line, which could be the x-axis, the y-axis or the equation of a graph.

For a reminder about equations of straight lines turn to page 47.

Now try this

1 Reflect the shape in the mirror line.

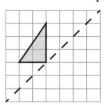

2 Describe fully the transformation that maps shape B onto shape C.

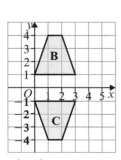

Rotation

A **rotation** is a type of transformation. You rotate a shape by turning it about a point called the **centre of rotation**.

To describe a rotation you need to give:

- the centre of rotation
- the angle of rotation
- the direction of rotation.

The centre of rotation may be the origin (0, 0) or another coordinate point.

For a reminder about coordinates turn to page 45.

The angle of rotation is given in degrees.

The direction of rotation will be clockwise or anticlockwise.

Rotated shapes are **congruent**.

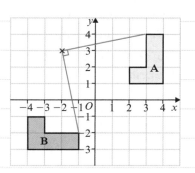

A to B: Rotation 90° clockwise about the point (−2, 3)

Worked example

On the grid, rotate shape A through 180° about point (0, 1).

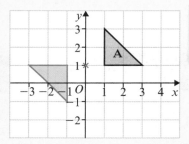

Everything in red is part of the answer.

Mark the centre of rotation (0, 1) with ✗.

Trace the shape.

Put your pencil on the cross.

Turn the tracing paper 180° to rotate the shape.

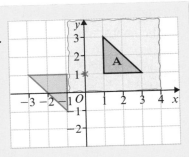

Worked example

Describe fully the single transformation that maps shape A onto shape B.

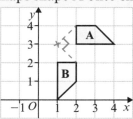

The transformation that maps shape A onto B is a rotation 90° clockwise about the point (1, 3)

To find the centre of rotation, trace the object shape then rotate the tracing paper, holding a point fixed with your pencil. Repeat for different points until your tracing covers the image.

Now try this

1 Rotate this shape 180° about the origin.

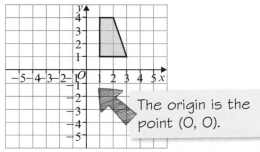

The origin is the point (0, 0).

2 Describe fully the single transformation that maps shape A onto shape B.

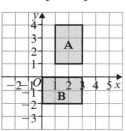

Translation

A **translation** is a type of sliding transformation.

You can use a **vector** to describe a translation.

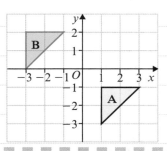

The transformation A → B is a translation by the vector

$$\begin{pmatrix} -4 \\ 3 \end{pmatrix}$$

The top number describes the horizontal movement.
- If the number is positive move right →
- If the number is negative move left ←

The bottom number describes the vertical movement
- If the number is positive move up ↕
- If the number is negative move down ↕

Translated shapes are **congruent**.

Worked example

Translate shape A by the vector $\begin{pmatrix} 2 \\ -1 \end{pmatrix}$. Label the new shape B.

Everything in red is part of the answer.

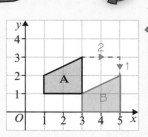

Move each vertex on shape A 2 squares right and 1 square down. Draw the image B.

Always read the question carefully to identify the object and the image.

When describing a translation you first need to write the word 'translation' and then use a vector to describe the translation.

$$\begin{pmatrix} \leftrightarrow \\ \updownarrow \end{pmatrix}$$ Movement left or right
         Movement up or down

Read the question carefully. You need to write the vector from A to B.

Worked example

Describe fully the single transformation that will map shape A onto shape B.

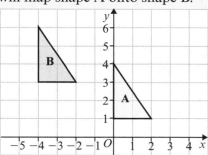

Translation by vector $\begin{pmatrix} -4 \\ 2 \end{pmatrix}$

Now try this

1 Describe fully the transformation that will map shape A onto shape B.

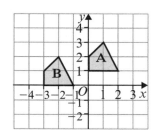

2 On the grid, translate shape S by the vector $\begin{pmatrix} 3 \\ -2 \end{pmatrix}$. Label the new shape T.

Enlargement

When you **enlarge** a shape you make it either bigger or smaller than the original shape.

The **scale factor** of an enlargement tells you how much to multiply each length by.

Lines drawn through corresponding points on the object A and the image B meet at the **centre of enlargement** X.

Enlarging a shape gives a **similar** shape. The angles in the shapes do not change but the lengths of the sides do change.

To find out more about similar shapes turn to page 86.

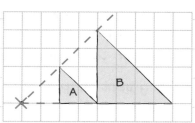

The scale factor from A to B is 2.

The scale factor from B to A is $\frac{1}{2}$.

Worked example

On the grid, enlarge the shape by scale factor 2 from the centre of enlargement X.

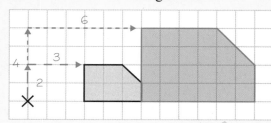

1. Count the squares horizontally and vertically from the centre of enlargement to each vertex on the shape.

2. Multiply all the distances from the centre by the scale factor.

3. The position of the top right vertex of the shape changes from 2 up and 3 right to 4 up and 6 right.

Check it!

Each length in the object should have been multiplied by the scale factor 2.

Object: base = 3
Image: base = 6 = 2 × 3 ✓

Worked example

Describe fully the single transformation that maps shape A onto shape B.

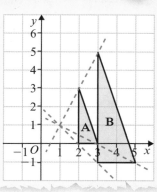

Enlargement, scale factor 2 with centre of enlargement (1, 1)

To fully describe this you need to state that it is an enlargement, and give the scale factor and the centre of enlargement.

Now try this

1 Enlarge this shape by scale factor 3 about the centre of enlargement X.

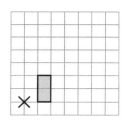

2 Describe fully the single transformation that maps shape A onto shape B.

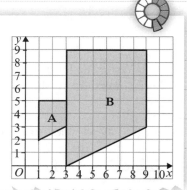

Congruent and similar shapes

Two shapes are **congruent** if they are exactly the same shape and size. Shapes are **similar** if one of the shapes is an enlargement of the other.

You need to be able to recognise congruent and similar shapes.

Congruent shapes are identical – their lengths are the same and their angles are the same size.

The shapes in each pair are congruent.

Reflected shapes

Rotated shapes

Translated shapes

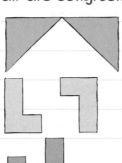

Similar shapes have **equal angles** and the sides are in the same **ratio**.

Enlargements produce similar shapes.

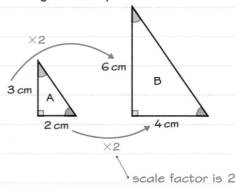

scale factor is 2

For a reminder about enlargement turn to page 85.

Worked example

The diagram shows six trapeziums.

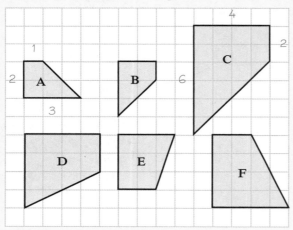

(a) Write the letter of the shape that is congruent to shape A.

B

(b) Write the letter of the shape that is similar to shape A.

C

(c) Write the letter of the shape that is congruent to shape F.

D

To test for congruency, use tracing paper to trace shape A and then see if your tracing paper fits over another shape.

Now try this

The diagram shows six triangles.

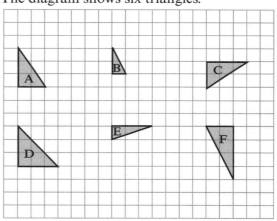

(a) Write the letters of one pair of congruent triangles.

(b) Write the letters of one pair of similar triangles.

Write down the lengths of the horizontal and vertical sides in shape A (1, 2 and 3). Find another shape with corresponding sides in the same ratio. The sides of shape C are 2, 4 and 6. Shape A is enlarged by scale factor 2 to give shape C.

Shape problem-solving

When you are solving shape problems, use what you know about shape as well as other maths topics such as algebra. You may not always answer a problem with one calculation. You may need to break the problem into several steps to find the final answer.

To revise solving equations look at page 39. To revise volume of cuboids look at page 65. For a reminder about angle facts look at page 69, and for area and perimeter of rectangles look at pages 75 and 76.

Worked example

DVD cases are cuboids measuring 15 cm by 20 cm by 2 cm.
The DVD cases are packed in larger cuboid boxes measuring 100 cm by 90 cm by 20 cm. The DVD cases are packed so that there are no gaps.
How many DVD cases can fit into one large box?

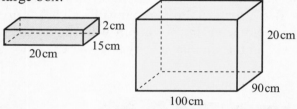

Volume of large box = 100 × 90 × 20
$\qquad$ = 180 000 cm³
Volume of DVD case = 20 × 15 × 2
$\qquad$ = 600 cm³
180 000 ÷ 600 = 300
300 cases

Problem solved!

The solution shown divides the volume of the large box by the volume of the DVD case. For this to be valid, you need to make sure that all the cases can be laid out in one layer without gaps, and that the layers will stack without gaps. The edge lengths of the large box must be multiples of the corresponding lengths of the DVD case.

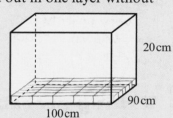

100 = 20 × 5 ✓ 90 = 15 × 6 ✓ 20 = 2 × 10 ✓
An alternative approach is to work out how many DVD cases fit in one layer (5 × 6 = 30), how many layers fit in the large box (20 ÷ 2 = 10), then multiply these (30 × 10 = 300).

> You'll need brilliant problem-solving skills to succeed in GCSE – get practising now!

Problem solved!

1. Write an equation, using the fact that angles in a triangle add up to 180°.
2. Solve the equation to find the value of x.
3. Remember to answer the question. Use your value for x to calculate the size of the other two angles. State the size of the largest angle.

Check it! Check that your angles add up to 180°.
25 + 75 + 80 = 180° ✓

> You'll need brilliant problem-solving skills to succeed in GCSE – get practising now!

Worked example

Calculate the size of the largest angle in this triangle.

Angles in a triangle add up to to 180°.
$2x + 30° + x + 3x = 180°$
$6x + 30° = 180°$
$6x = 150°$
$x = 25°$
$3x = 3 × 25 = 75°$
$2x + 30° = 2 × 25 + 30 = 80°$
The largest angle is 80°

You need to find the value of x. Set up an equation using the information that the perimeter is 40 cm. So $40 = 3x + 3x + x + x$. Solve the equation and use your value for x in the formula for the area of a rectangle: area = length × width. Don't forget to include the units for area in your answer.

Now try this

The perimeter of this rectangle is 40 cm.
Calculate the area of the rectangle.

Probability

Probability is a measure of how likely an event is to happen.

All probabilities have a value between 0 and 1.

You can use fractions, decimals and percentages to describe probabilities.

An event that is **certain** has a probability of 1 or 100%.

An event that is **impossible** has a probability of 0 or 0%.

Impossible	Even chance	Certain
0	$\frac{1}{2}$	1
	0.5	
0%	50%	100%

It is **very likely** that it will rain in the next 2 months so A is close to 1.

There is an **even chance** that when you flip a coin it will land heads up, so B is halfway.

It is **very unlikely** that someone in your family will win the jackpot next week so the probability of C is close to 0.

Worked example

On this probability scale mark the probability of each event:

```
C                    B                    A
├────────────────────┼────────────────────┤
0                                          1
```

A   It will rain some time in the next 2 months.
B   When you flip a coin it will land heads up.
C   Someone in your family will win the jackpot in the national lottery next week.

Writing probabilities

You can write the probability of an event happening using the notation P(event).

When you flip a coin the probability of it landing heads up is $\frac{1}{2}$. You write this as P(Heads) or P(H) = $\frac{1}{2}$

There is only one head. There are two possible outcomes: heads and tails.

Worked example

A normal six-sided dice is rolled. Calculate
(a) P(3)   (b) P(odd number)
(a) P(3) = $\frac{1}{6}$ (b) P(odd) = $\frac{3}{6}$

The possible outcomes when you roll a dice are: 1, 2, 3, 4, 5, 6
There is only one 3 so P(3) = $\frac{1}{6}$

Golden rule

$P(\text{event}) = \dfrac{\text{number of successful outcomes}}{\text{total number of possible outcomes}}$

Worked example

The diagram shows a normal eight-sided spinner.
(a) Work out the probability this spinner will land on blue or green.
P(blue or green) = $\frac{3}{8}$
(b) Work out the probability the spinner will land on yellow.
P(yellow) = 0

(a) Three sections are blue or green. There are eight possible outcomes.

(b) There are no yellow sections so it is impossible for the spinner to land on a yellow.

Now try this

List all the possible outcomes.

The diagram shows a normal six-sided spinner.
     (a)   Which letter is the spinner most likely to land on?
     (b)   Work out the probability that the spinner lands on A.

Outcomes

You can calculate probabilities by listing all the **outcomes** of an event.

When you flip a coin, the only possible outcomes are heads and tails.

There are two different outcomes and each is **equally likely** to occur.

P(Head) = $\frac{1}{2}$ P(Tail) = $\frac{1}{2}$ P(H) + P(T) = 1

The probabilities of all the different possible outcomes of an event add up to 1.

Worked example

Jon flips two coins. What is the probability of both of them landing heads up?

Possible outcomes: HH, HT, TH, TT

P(HH) = $\frac{1}{4}$

Write all the possible outcomes.

There are four possible outcomes.

Only one is HH.

Use the formula

$$P(\text{event}) = \frac{\text{number of successful outcomes}}{\text{total number of possible outcomes}}$$

Golden rule

If you know the probability that something will happen, you can calculate the probability that it will **not** happen.

P(outcome doesn't happen) = 1 − P(outcome happens)

The probability of this spinner landing on blue is $\frac{1}{6}$ so the probability of it **not** landing on blue is $1 - \frac{1}{6} = \frac{5}{6}$

If the probability is given as a decimal, write the answer as a decimal.

If it is written as a fraction, write your answer as a fraction.

Check your answer.
$\frac{1}{4} + \frac{3}{4} = \frac{4}{4} = 1$ ✓

Worked example

(a) The probability that it will rain tomorrow is 0.6. Work out the probability that it will not rain tomorrow.

P(Will not rain) = 1 − P(will rain)
              = 1 − 0.6 = 0.4

(b) The probability that Josie is late for school is $\frac{1}{4}$. What is the probability that she is not late for school?

P(Josie is not late) = 1 − P(Josie is late)
              = 1 − $\frac{1}{4}$ = $\frac{3}{4}$

Now try this

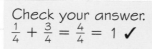

1   The probability that Mark wins a game of tennis is 0.8
What is the probability he doesn't win?

2   The probability that a train is late is 30%.
What is the probability it is on time?

> Probabilities add up to 1 or 100%.

3   Carmen flips a coin and spins this spinner.

(a) Write a list of all the possible outcomes.

(b) Work out the probability of getting a tails on the coin and a 1 on the spinner.

> Be systematic in writing out all the possible outcomes.

Experimental probability

You can use the results of an experiment to estimate probabilities. This is called the **experimental probability**. The more trials you do, the more reliable the probability.

Expected outcome is an estimate of how many times you expect something to happen. For example, when you flip a fair coin 100 times, you expect to get a tails about 50 times.

Use these formulae to calculate the probabilities and outcomes of experiments.

1 $\dfrac{\text{Experimental}}{\text{probability}} = \dfrac{\text{frequency of outcome}}{\text{total frequency}}$

2 $\dfrac{\text{Expected}}{\text{outcome}} = \dfrac{\text{number of}}{\text{trials}} \times \dfrac{\text{probability of}}{\text{successful outcome}}$

Theoretical probability

Theoretical probability is calculated without doing an experiment. For example, the probability of rolling a 3 when you roll a fair dice is $\frac{1}{6}$

If you do an experiment rolling a fair dice, the more trials you do, the closer the experimental probability will get to the theoretical probability.

Worked example

Fernando rolls a dice 100 times. He records the results in a table.

(a) Find the experimental probability of rolling each number.

Number on dice	Frequency	Experimental probability
1	21	$\frac{21}{100}$
2	13	$\frac{13}{100}$
3	8	$\frac{8}{100}$
4	10	$\frac{10}{100}$
5	11	$\frac{11}{100}$
6	37	$\frac{37}{100}$

(b) Do you think this is a fair dice?
No. If it were fair, you would expect similar frequencies. There are a lot more 6s than expected. The dice is probably biased.

Worked example

The probability of a spinner landing on yellow is 0.2
The spinner is spun 200 times.
Work out an estimate for the number of times the spinner will land on yellow.
$200 \times 0.2 = 40$

There are 200 trials and the probability of the spinner landing on yellow is 0.2

Now try this

1 A bag contains an unknown number of coloured counters.

   Pedro picks a counter at random from the bag, notes its colour, then replaces it. He does this 50 times and his results are recorded in the table below.

Colour	Red	Green	Yellow	Pink
Frequency	12	14	20	4

   (a) What is the probability that the next counter Pedro picks is green?

   (b) How many times would Pedro expect to pick a green counter out of the bag if he repeated the experiment 200 times?

   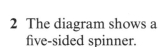
   P(Green) × 200

2 The diagram shows a five-sided spinner.

   Andrea spins the spinner and records the probability of spinning each number in a table.

   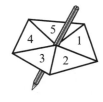

Number	1	2	3	4	5
Probability	0.2	0.3	0.1	0.1	0.3

   She spins the spinner 500 times. Work out an estimate for the number of times the spinner will land on 1.

Probability diagrams

A **sample space diagram** shows all the possible outcomes of two events.
Here are all the possible outcomes when you flip two coins. There are
four possible outcomes altogether: HH, HT, TH and TT.
Other diagrams that are useful are **two-way tables** and **Venn diagrams**.

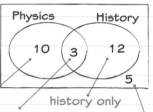

Second coin

First coin	H	T
H	HH	HT
T	TH	TT

TH means getting a tails on the first coin and a heads on the second coin.

Worked example

Joanne flips a coin and rolls a fair six-sided dice.
(a) Draw a sample space diagram to show all
the possible outcomes.

Dice

Coin	1	2	3	4	5	6
H	H1	H2	H3	H4	H5	H6
T	T1	T2	T3	T4	T5	T6

(b) Work out the probability that Joanne rolls
an odd number and flips heads.

P(Odd, Heads) = $\frac{3}{12}$ = $\frac{1}{4}$

There are 3 outcomes for odd number and
heads: H1, H3 and H5, so the numerator is 3.
There are 12 possible outcomes altogether, so
the denominator is 12.

Venn diagrams

This Venn diagram
shows the numbers
of students who
study physics and
history in a class
of 30 students.

Physics   History

10   3   12

5

physics only   history only
both physics   neither history
and history   nor physics

The **number** in each section tells you **how
many** people that section represents.
A total of 10 + 3 + 12 + 5 = 30 students

If a student is picked from the class at
random, P(Physics and history) = $\frac{3}{30}$

Worked example

Jamie did a survey of the eye colour and hair
colour of 14 of his friends.
2 friends had brown hair and blue eyes.
7 friends had brown hair.
9 friends had blue eyes.
(a) Draw a Venn diagram of this information.

Brown hair   Blue eyes

5   2   7

Always start with
the intersection.
7 − 2 = 5
9 − 2 = 7

(b) Work out the probability that one of these
friends had brown hair but not blue eyes.

P(Brown hair and not blue eyes) = $\frac{5}{14}$

Worked example

28 students were asked if they preferred tennis
or badminton.
8 boys said they preferred tennis.
Of the 11 students who preferred badminton,
5 were girls.
(a) Complete this two-way table.

	Tennis	Badminton	Total
Boys	8	6	14
Girls	9	5	14
Total	17	11	28

(b) What is the probability that a student
picked at random preferred tennis?

P(Preferred tennis) = $\frac{17}{28}$

Now try this

Claire rolls two dice and adds the results together.
(a) Complete this sample space diagram.
(b) How many possible outcomes are there?
(c) Which total is most likely?
(d) Work out the probability that the total of the two
dice is 10.

First dice

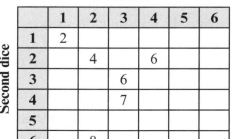

Second dice	1	2	3	4	5	6
1	2					
2		4		6		
3			6			
4			7			
5						
6		8				

Probability tree diagrams

A **tree diagram** shows all the outcomes of two or more events.

You work out the probability of different outcomes by multiplying along the branches.

Probabilities on tree diagrams

The probability of Jamie being late for school on any day is 0.1

The tree diagram shows the probability of Jamie being late for school two days in a row.

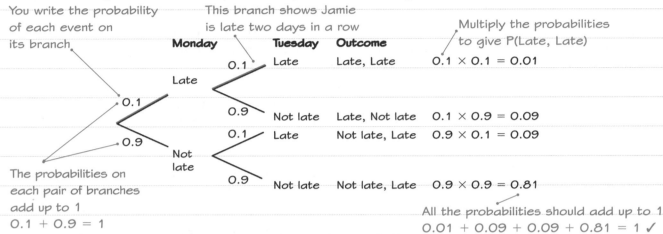

You write the probability of each event on its branch.

This branch shows Jamie is late two days in a row

Multiply the probabilities to give P(Late, Late)

Monday	Tuesday	Outcome	
	0.1 Late	Late, Late	0.1 × 0.1 = 0.01
Late 0.1	0.9 Not late	Late, Not late	0.1 × 0.9 = 0.09
Not late 0.9	0.1 Late	Not late, Late	0.9 × 0.1 = 0.09
	0.9 Not late	Not late, Late	0.9 × 0.9 = 0.81

The probabilities on each pair of branches add up to 1
0.1 + 0.9 = 1

All the probabilities should add up to 1
0.01 + 0.09 + 0.09 + 0.81 = 1 ✓

P(Jamie is late two days in a row) = 0.1 × 0.1 = 0.01

Worked example

Marcus plays tennis. He either wins a game or doesn't win.
The probability that Marcus wins a game is $\frac{3}{4}$.
He plays two games of tennis.
(a) What is the probability Marcus doesn't win a game?

P(Doesn't win) = 1 − P(Wins)
   = 1 − $\frac{3}{4}$ = $\frac{1}{4}$

(b) Complete this probability tree diagram to show the probabilities of all the possible outcomes.

(c) Work out the probability that Marcus wins both games.

P(Wins, Wins) = $\frac{3}{4} \times \frac{3}{4} = \frac{9}{16}$

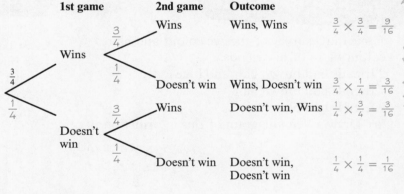

1st game	2nd game	Outcome	
	Wins $\frac{3}{4}$	Wins, Wins	$\frac{3}{4} \times \frac{3}{4} = \frac{9}{16}$
Wins $\frac{3}{4}$	Doesn't win $\frac{1}{4}$	Wins, Doesn't win	$\frac{3}{4} \times \frac{1}{4} = \frac{3}{16}$
Doesn't win $\frac{1}{4}$	Wins $\frac{3}{4}$	Doesn't win, Wins	$\frac{1}{4} \times \frac{3}{4} = \frac{3}{16}$
	Doesn't win $\frac{1}{4}$	Doesn't win, Doesn't win	$\frac{1}{4} \times \frac{1}{4} = \frac{1}{16}$

Now try this

The probability that Peter passes a maths test is $\frac{1}{2}$

The probability that Josie passes the test is $\frac{3}{4}$

(a) What is the probability Peter does not pass the test?

(b) What is the probability that Josie does not pass the test?

(c) Complete this tree diagram.

(d) Work out the probability that both Peter and Josie pass the maths test.

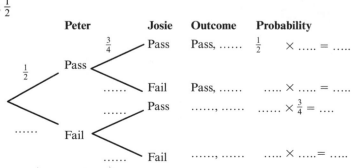

Peter	Josie	Outcome	Probability
	Pass $\frac{3}{4}$	Pass, ...	$\frac{1}{2}$ × ..... = ...
Pass $\frac{1}{2}$	Fail	Pass, ...	..... × ..... = ...
Fail	Pass	......, ...	..... × $\frac{3}{4}$ = ...
	Fail	......, ...	..... × ..... = ...

Mutually exclusive and independent events

You need to be able to identify when events are independent or mutually exclusive.

 1 Two events are **mutually exclusive** if they cannot happen at the same time. For example, when you flip a coin once, you cannot get a heads and a tails at the same.

2 **Independent events** are events that do not affect each other. So the probability of one event **does not affect** the probability of the other event. If you flip a coin and roll a dice at the same time, whether you get tails doesn't affect whether you get a 3.

Events that are NOT independent are called **dependent events**.

Worked example

Jan rolls an ordinary six-sided dice once. Which of these events are mutually exclusive?
(a) Rolling a multiple of 3 and rolling an even number

Not mutually exclusive because 6 is both a multiple of 3 and an even number.

(b) Rolling an odd number and rolling a multiple of 4

Mutually exclusive because no odd number is also a multiple of 4.

Venn diagrams are useful in showing whether two events are mutually exclusive or not.

Even number | Multiple of 3
2 4   6   3
5   1

Odd number | Multiple of 4
1 3   5   4
2   6

There can't be anything in the overlap if events are mutually exclusive.

(a) These are independent events because for example, the probability of rolling a 4 on the dice does not affect the probability of getting a tails on the coin.

(b) These are not independent events because picking and eating a chocolate means there are fewer chocolates left, so the probability of picking another chocolate will be lower.

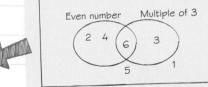

Worked example

Which of these events are independent events?
(a) Rolling a dice and then flipping a coin
Independent events
(b) From a bag of sweets containing toffees and chocolates, picking a chocolate at random, eating it and then picking another chocolate
Dependent events

Now try this

Jason rolls a fair six-sided dice.

Which of these events are mutually exclusive? Give your reasons.

Use a Venn diagram to show whether events are mutually exclusive.

(a) Rolling a 5 and an even number

(b) Rolling a multiple of 3 and a number smaller than 2

(c) Rolling an odd number and a square number

Averages and range

There are three different types of average: **mean**, **median** and **mode**. Averages tell you a typical value of a set of data. The range tells you how spread out the data set is.

Worked example

Here is a set of data.

6   7   9   3   9   9   8   13

(a) Write the mode.

The mode is 9.

>  The mode is the value that occurs **most often**.

(b) Work out the mean.

6 + 7 + 9 + 3 + 9 + 9 + 8 + 13 = 64

64 ÷ 8 = 8

The mean is 8.

> To find the mean, add up all the values and divide by the number of values.

(c) Work out the median.

3̶ 6̶ 7̶ 8 9 9̶ 9̶ 1̶3̶

Median = 8.5

> The median is the **middle value**. Write the numbers in order, from the smallest value to the largest value. Cross off the smallest and largest values, in pairs, until you reach the middle.
>
> If there are two numbers in the middle, the median is halfway between them.

(d) Work out the range.

13 − 3 = 10

The range is 10.

> Range = biggest value − smallest value

Worked example

The table shows the numbers of merits awarded in one month to a class.

Number of merits	Frequency
0	2
1	3
2	5
3	7
4	3
5	3

(a) How many students were in the class?

2 + 3 + 5 + 7 + 3 + 3 = 23

(b) What is the modal number of merits?

The modal number of merits is 3.

(c) What is the median number of merits?

The median number of merits is 3.

median
11   ↑   11
       12th value

(b) The modal number is the same as the mode.

(c) There were 23 merits altogether.

The median is the $\frac{23 + 1}{2}$ = 12th value. Find which frequency group the 12th value is in. Remember that the answer is the number of merits, not the frequency.

Worked example

Here are three number cards.

? ? ?

The mode of the numbers on the cards is 7.
The mean of the three numbers is 6.
Work out the three numbers on the cards.

The numbers on the three cards are 4, 7, 7

Problem solved!

1. The mode is 7 so at least two cards must be 7s.
2. The mean is 6 so the **sum of the numbers** is 6 × 3 = 18
   So 7 + 7 + ? = 18 → the other number must be 4
3. **Check it!**
   7 + 7 + 4 = 18 ✓   18 ÷ 3 = 6 ✓

> You'll need brilliant problem-solving skills to succeed in GCSE – get practising now! 💡

Now try this

Start with the median.

Here are three different number cards.

? ? ?

The median, mean and range are all 4.
Work out the number on each card.

Averages from tables

When you have lots of data it may be useful to use a frequency table to display the data. You can find averages from **ungrouped** and **grouped frequency tables**.

Worked example

The table shows the numbers of text messages sent by some students one lunchtime.

Everything in red is part of the answer.

Number of text messages x	Frequency f	Frequency × number of texts ($f \times x$)
1	6	$1 \times 6 = 6$
2	5	$2 \times 5 = 10$
3	7	$3 \times 7 = 21$
4	2	$4 \times 2 = 8$

(a) Write the mode.

The mode is 3 text messages.

(b) Work out the median.

The median is halfway between the 10th and 11th values. The median is 2 text messages.

(c) Work out the mean.

Total of ($f \times x$) column is $6 + 10 + 21 + 8 = 45$

Total frequency is $6 + 5 + 7 + 1 + 2 = 20$

$45 \div 20 = 2.25$

The mean is 2.25 text messages.

To calculate the mean from a frequency table you need to add an extra column, Frequency × number of texts, $f \times x$. The total of this column is the total number of texts made. Use this rule to work out the mean: Mean = $\dfrac{\text{total of } (f \times x) \text{ column}}{\text{total frequency}}$

Worked example

The table shows the times it took a class of students to travel to school. Calculate an estimate for the mean time taken to travel to school.

Everything in red is part of the answer.

Time, t (mins)	Frequency	Midpoint	($f \times x$)
$0 < t \le 10$	0	5	$5 \times 0 = 0$
$10 < t \le 20$	5	15	$15 \times 5 = 75$
$20 < t \le 30$	12	25	$25 \times 12 = 300$
$30 < t \le 40$	2	35	$35 \times 2 = 70$
$40 < t \le 50$	1	45	$45 \times 1 = 45$
	Total $f = 20$		Total $f \times x = 490$

$$\text{Mean} = \frac{\text{total } f \times x}{\text{total } f}$$
$$= \frac{490}{20}$$
$$= 24.5 \text{ minutes}$$

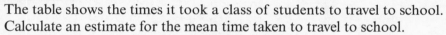

As you don't have an exact value for the time, you have to use an **estimate**. The midpoint is a good estimate. Add another column and find the **midpoint** of each group. The midpoint of 0 and 10 is 5.

Now try this

The table shows the heights, h cm, of 12 plants.

Height, h (cm)	Frequency, f	Midpoint, x	Frequency × midpoint ($f \times x$)
$0 < h \le 20$	2		
$20 < h \le 40$	3		
$40 < h \le 60$	5		
$60 < h \le 80$	2		
	Total frequency =		Total $f \times x$ =

(a) Complete the two columns in the table.

(b) Calculate an estimate for the mean height of the plants.

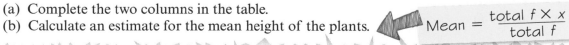 Mean = $\dfrac{\text{total } f \times x}{\text{total } f}$

Stem and leaf diagrams

An ordered stem and leaf diagram is made up of a 'stem' and a 'leaf' and the data are arranged in order of size.

This stem and leaf diagram shows the marks of 19 students in a spelling test.

0|8 represents 8
The smallest value is 8.

stem	leaf
0	8 9
1	3 6 7
2	4 4 4 5 6 8 9
3	1 2 4 4 5 5 8

The largest value is 38.

Key: 1|6 = 16 marks

The key is important because it tells you about the value of the data in the stem and leaf diagram.

You can use stem and leaf diagrams to find averages.

For a reminder about averages turn to page 94.

Worked example

Pedro recorded the numbers of spam emails his friends received in one day. These are his results.

23, 5, 12, 4, 14, 14, 24, 35, 27, 6,
12, 14, 28, 18, 32

(a) Show these data in an ordered stem and leaf diagram.

(a)
0	5 4 6
1	2 4 4 2 4 8
2	3 4 7 8
3	5 2

0	4 5 6
1	2 2 4 4 4 8
2	3 4 7 8
3	2 5

Key: 1 | 2 = 12 emails

(b) Work out the range.
Range = 35 − 4 = 31 spam emails
(c) Write the mode.
Mode = 14 spam emails
(d) Work out the median.
Median = 14 spam emails

To draw a stem and leaf diagram:

1. Look at the smallest and largest values in the data – these will help you choose sensible values for your stem.

2. Draw an ordered stem.

3. Cross off each data value as you enter it into an unordered stem and leaf diagram.

4. Draw another diagram, putting the data (the leaves) in order.

5. Write a key.

The mode is the most common value.

The median is the middle value when the data are in order from smallest to largest. Cross off the data values in pairs going forward and backwards, until you reach the middle value.

Now try this

Here are the times taken, in minutes, to complete a puzzle.

23, 41, 37, 22, 43, 34, 38, 23,

33, 23, 34, 21, 42, 34, 35, 34

(a) Show these data in an ordered stem and leaf diagram.

(b) Work out the range.

(c) Write the mode.

(d) Work out the median.

Start by crossing off the lowest and highest values. The median is the middle value. When there is an even number of data values, the median is halfway between the two middle values.

Analysing data

In statistics, a **population** is a group you are interested in. A whole population is usually too big to collect data about, so you choose a **sample**, which is a smaller group chosen from the population.

- The sample must be large enough to be reliable – about 10% is a good sample.

- The sample must be **random**. Every member of the population must have an equal chance of being included.

Hypothesis

A **hypothesis** is an idea you want to test. When testing a hypothesis, make sure you collect the data you need.

Worked example

Jo does a survey about how students in her school travel to school. There are 800 students in her school. Choose the most appropriate sample size for her survey.

A 8 students
B 80 students
C 800 students

B 80 students

8 students is not a good sample size because there are too few students.

80 is 10% of 800 so it is a good sample size.

800 is all of the students in the school or population. It would take a long time to ask everybody – that is why you should use a **sample**.

Worked example

Jamie wants to test the hypothesis: 'Students who travel by bus are often late for school.' Which of these sets of data does he need to collect?

A Age of students
B How often students are late
C Gender of students
D Method of transport

B and D

The age and gender of students are not relevant to the hypothesis being tested.

Worked example

Hilary wants to find out the numbers of adults, boys and girls who visit a museum.
Design a suitable data collection sheet to collect the information.

	Tally	Frequency
Adults		
Boys		
Girls		

A random sample is a good way to collect data but it is important to collect data from a random sample in an unbiased way.

Now try this

Claire wants to do a survey to find out the most popular sport in her school.

There are 500 students in her school.

(a) How many students should she sample?

(b) Which of these methods of collection would affect the data? Give your reasons.

A A random sample of students in her year
B A random sample of students in the school
C Students in the football and netball teams
D A random sample of boys

Pie charts

A **pie chart** is a circle divided into slices called sectors.

The whole circle represents a set of data.

Each sector represents a fraction of the data.

This pie chart shows the colours of cars in a car park.

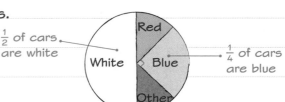

$\frac{1}{2}$ of cars are white

$\frac{1}{4}$ of cars are blue

Worked example

Kami asked 24 students about their favourite drinks. The table shows her results.

Draw a pie chart to show this information.

Drink	Number of students	Angle
Orange	8	8 × 15° = 120°
Blackcurrant	6	6 × 15° = 90°
Lemon	6	6 × 15° = 90°
Other	4	4 × 15° = 60°

Total number of students = 8 + 6 + 6 + 4 = 24

Angle for 1 student = 360° ÷ 24 = 15°

Check: 120° + 90° + 90° + 60° = 360° ✓

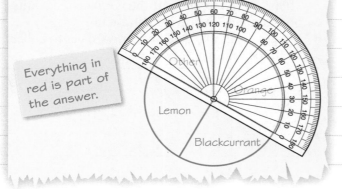

Everything in red is part of the answer.

You need a sharp pencil, compasses and a protractor to draw a pie chart.

1. Add a new column to the table and label it 'Angle'.

2. Work out how many degrees will represent 1 student.

   There are 360° in a circle and 24 students so the number of degrees for 1 student is 360° ÷ 24 = 15°.

3. Multiply the number of students by the number of degrees for 1 student to give you the angle for each sector.

4. Check that all your angles add up to 360°.

5. Draw a circle. Draw a vertical line and use a protractor to measure the first angle (120°) from this line. Draw the rest of the angles, taking care not to overlap the angles. It helps to turn your page so that you measure from the last angle you drew.

6. Label each sector of your pie chart.

For a reminder about measuring and drawing angles turn to page 67.

Now try this

The table shows information about the members of a chess club.

Members	Frequency	Angle
Boys	15	
Girls	10	
Adults	11	

Angle for 1 person = 360° ÷ number of people

(a) How many people are members of the chess club?

(b) Work out the number of degrees that represents one person.

(c) Complete the angle column in the table.

(d) Draw a pie chart to show this information.

Scatter graphs

A **scatter graph** compares two sets of data on the same graph.

The shape of the scatter graph shows if there is a relationship between the two sets of data.

If the points in a scatter graph lie roughly on a straight line then the scatter graph shows **correlation**.

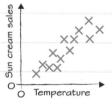

Positive correlation

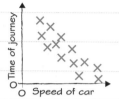

Negative correlation

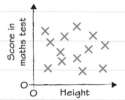

No correlation

Worked example

The scatter graph shows the marks obtained by 10 students in two maths tests.

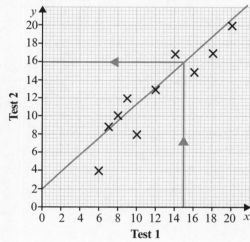

Test 1

(a) What type of correlation does the scatter graph show?

Positive correlation

(b) Draw a line of best fit on the scatter graph.

(c) Another student scored 15 marks on test 1. Use your line of best fit to estimate the student's mark on test 2.

16 marks

As the marks in test 1 increase the marks in test 2 also increase. This is positive correlation.

The **line of best fit** is a straight line that is as close to the points as possible.

You can use your line of best fit to predict values that are not plotted.

Describe what happens to the rainfall as the temperature increases.

Now try this

The scatter graph gives information about the rainfall (mm) and temperature (°C) for 12 days in a year.

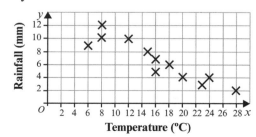

Temperature (°C)

(a) Describe the relationship between the temperature and rainfall.

(b) What type of correlation does the scatter graph show?

(c) Draw a line of best fit.

(d) On one day during the year the temperature was 19 °C. Use your line of best fit to estimate the rainfall on that day.

Writing a report

When you have completed a survey, it is useful to draw conclusions and write a report on what you have found out. You can calculate averages and use graphs and charts to support your conclusions and show your results in a clear and organised way.

Golden rules

A report should include:

- the hypothesis that you are testing
- averages and range
- at least one graph or chart
- a conclusion
- ideas on how to improve the investigation.

Worked example

Josie wants to find out if students own more than 1 pet. She asks 30 students in Year 9. The table shows the numbers of pets these students own.

Write a report to clearly show Josie's findings. Include a pie chart in your report.

Number of pets	Frequency	Number of pets X frequency	Angle
0	2	0	2 × 12° = 24°
1	6	6	6 × 12° = 72°
2	9	18	9 × 12° = 108°
3	8	24	8 × 12° = 96°
4	5	20	5 × 12° = 60°
	30	68	

Mode = 2 pets   Range = 4 − 0 = 4 pets
Mean = total frequency ÷ total number of pets = 68 ÷ 30 = 2.3 pets
Angle for 1 student = 360° ÷ 30 = 12°
Check: 24° + 72° + 108° + 96° + 60° = 360°

The modal number of pets is 2 and the mean number of pets is 2.3. The range is 4 pets. As the average number of pets owned is 2 or 2.3, depending on whether you use the mode or the mean, Josie has shown that, on average, people own more than 1 pet.

To improve the investigation, Josie could survey different Year groups to find out if older students owned more pets than younger students.

Number of pets

Now try this

A scientist wanted to test which of two different soil types was better for growing plants. She planted seeds in two different soil samples, A and B, and measured the heights of the plants (cm) after 4 weeks.

Soil A (cm)	5	15	25	32	4	35	17	25	25	30
Soil B (cm)	10	20	24	10	20	2	23	13	25	38

(a) Work out the mean, median, mode and range for the heights of plants grown in Soil A and in Soil B.

(b) Draw a stem and leaf diagram for the heights of the plants grown in each of the soil samples.

(c) Write a conclusion. Write an explanation of which soil sample you think is better for growing plants.

Answers

NUMBER

1. Whole numbers
1. (a) Ninety-five thousand, three hundred and sixty-four
   (b) 5000 or five thousand
2. (a) 3756, 3765, 20 098, 21 089, 21 465, 1 000 010
   (b) $-11, -3, -1, 4, 6, 9$
3. Moscow

2. Decimals
1. (a) $6.7 > 6.456$ (b) $23.819 < 23.84$
2. 2.15, 2.199, 2.26, 2.3
3. 6 hundredths or $\frac{6}{100}$
4. Kris's

3. Rounding
1. $4000 + 3000 = 7000$ with both numbers rounded up, but both are less than their rounded number so $3790 + 2858 < 7000$
2. $6 \times 3 = 18$
3. (a) 3800   (b) 25   (c) 0.74

4. Addition
1. 8464
2. 49.025
3. 16.363 m
4. Lukas; Oliver spent £56.54 and Lukas spent £56.65

5. Subtraction
1. 919
2. 0.65 kg
3. £7.26

6. Understanding powers of 10
1. (a) 6.18   (b) 57.8   (c) 8.3   (d) 3700
2. 200 000
3. £0.45 million

7. Multiplication
1. (a) 19 992   (b) 25 848   (c) 9.1   (d) 114.048
2. 8520
3. £12
4. Rounding: $12 \times 5 = 60$, so the answer is incorrect.

8. Division of whole numbers
1. (a) 646   (b) 264
2. 32
3. 432, 468

9. Division with decimals
1. (a) 183.8 (b) 435
2. £35.64
3. 15

10. Negative numbers
1. (a) 1   (b) -6 (c) -9 (d) -5
2. (a) 12   (b) -36 (c) 6   (d) -5
3. Max is correct because negative $\times$ negative = positive, and then negative $\times$ positive = negative
4. $2\,°C$

11. Factors, multiples and primes
1. Diagram should show factors of 30: 1, 2, 3, 5, 6, 10, 15, 30
2. (a) (i) 1, 2, 4, 8, 16, 32
       (ii) 1, 2, 3, 4, 6, 8, 12, 16, 24, 48
   (b) 16
3. 56
4. (a) 23 or 29
   (b) Various numbers, e.g. 11, 31, 61

12. Squares, cubes and roots
1. (a) -8 (b) 7   (c) 10
2. (a) 3375   (b) 10.95 (2 d.p.)   (c) -18
3. $52\,m - 48\,m = 4\,m$

13. Priority of operations
1. (a) (i) 9   (ii) 16
   (b) (i) 86   (ii) 81
   (c) (i) 11   (ii) 19
2. $30^2 = 900; 25 \times 36 = 900$

14. More powers
1. (a) 4^6 (b) 8^8 (c) 10^3 (d) 10^5
2. (a) $\frac{1}{25}$ (b) $\frac{1}{9}$ (c) 1
3. (a) 625   (b) -216 (c) 0.008

15. Prime factors
1. $2^2 \times 3^2 \times 5$
2. (a) $2^3 \times 3$ (b) 2^6 (c) $2^2 \times 5 \times 7$ (d) $2^3 \times 5^2$

16. HCF and LCM
1. HCF 4; LCM 288
2. HCF 24; LCM 360
3. 16 m
4. 3 trays of small rolls and 4 trays of large rolls

17. Standard form
1. (a) 7500   (b) 0.064
2. (a) 5.6×10^3 (b) 9.9×10^{-2}
3. 3.4×10^5

18. Calculator buttons
1. (a) 0.35   (b) $\frac{11}{4}$ (c) $5\frac{1}{4}$
2. (a) $\frac{4}{5 - \frac{2}{3}}$ (b) $\frac{2}{15}$

19. Fraction basics
1. (a) $\frac{1}{3}$ (b) $\frac{1}{4}$
2. $\frac{1}{2}, \frac{7}{10}, \frac{15}{20}, \frac{4}{5}$
3. £125, £100, £25

20. Changing fractions

1. $\frac{50}{9}$
2. $7\frac{1}{2}$
3. $\frac{21}{5}$
4. 33 pieces

21. Add and subtract fractions

1. $1\frac{2}{3}$
2. $8\frac{7}{30}$
3. $\frac{11}{15}$ m

22. Multiply and divide fractions

1. $\frac{11}{16}$
2. 15
3. $\frac{4}{5}$ m

23. Fractions, division, decimals

1. (a) $0.\dot{4}2857\dot{1}$ (b) $0.41\dot{6}$
2. Two written division calculations
3. Yes: $\frac{2}{9} = 0.\dot{2}$, $\frac{1}{3} = 0.\dot{3}$, $\frac{1}{6} = 0.1\dot{6}$, $\frac{7}{12} = 0.583\dot{3}$
   ($\frac{3}{4} = 0.75$, $\frac{5}{8} = 0.625$)

24. Equivalence

1. $\frac{21}{25} = \frac{84}{100} = 84\%$; $\frac{17}{20} = \frac{85}{100} = 85\%$; 17 out of 20 is higher
2. $\frac{1}{5}$, 24%, $\frac{1}{4}$, 0.26, 6 out of 20

25. Percentages

£320, £192, £96, £32

26. Number problem-solving

1. £21.41
2. $43 \div 3 = 14.\dot{3}$ so they should each pay £14.$\dot{3}$ but this is not possible, so they need to pay £13.34 each to cover the bill (or $2 \times £14.33$ and $1 \times £14.34$)

ALGEBRA

27. Collecting like terms

1. (a) (i) $4e$ (ii) $2z$
   (b) (i) $7j + 3k$ (ii) $5 - 6a$
2. $6x + 8y$

28. Simplifying expressions

1. (a) $20ab$ (b) $-30ef$
2. (a) $6b$ (b) $5x^2$
3. $20xy\,\text{cm}^2$

29. Writing expressions

1. $15xy$
2. $6x - 7$
3. $200 - 5s$

30. Indices

1. (a) n^{10} (b) m^3 (c) k^{18}
2. $\frac{1}{64}$
3. (a) $-12a^5$ (b) $4m^2$ (c) 20
4. y^0 (or 1)
5. $\frac{y^{4-}}{y^{1+2}} = \frac{y^3}{y^3} = y^{3-3} = y^0 = 1$

31. Expanding brackets

1. (a) $4h - 20g$ (b) $5cd + 6c$
2. $15x^2 - 20xy + 40x$
3. $14a^2 + 14ab$
4. LHS $= 10e^2 - 20ef =$ RHS

32. Expanding double brackets

1. (a) $x^2 - 4x - 12$
   (b) $x^2 - 13x + 40$
   (c) $x^2 - 25$
   (d) $x^2 + 8x + 16$
2. $x^2 - 4$

33. Factorising

1. (a) $9(2y - 3q)$ (b) $p(7q - 5t)$
2. $7d^3(3 - 5d^2)$

34. Substitution

1. 55
2. 40
3. 6

35. Linear sequences

(a) 68
(b) -9
(c) 32, 23, 14
(d) Yes
(e) -4

36. The nth term

1. 1, 5, 9, 13, 17
2. (a) $4n - 1$ (b) $-2n + 6$

37. Non-linear sequences

1. 2, 5, 10, 17, 26
2. $n^2 + 5$
3. arithmetic: 100, 91, 82, 73, … and 2, 13, 24, 35, …
   geometric: 4, 8, 16, 32, … and 1, 5, 25, 125, …
   quadratic: 0, 3, 8, 15, 24, … and 25, 36, 49, 64

38. Writing equations

(a) $5n - 7 = 18$
(b) $8s = 32$
(c) $9t + 1 = 64$

39. Solving simple equations

1. (a) $x = 7$ (b) $p = 9$ (c) $b = 125$ (d) $a = 40$
2. (a) $x = 7$ (b) $y = 3$ (c) $a = 20$ (d) $b = 28$

40. Solving harder equations

1. (a) $x = 3$ (b) $x = 6$ (c) $y = -2$ (d) $x = 5$
2. 5

41. Inequalities

1. $-1 \leqslant x \leqslant 3$
2. $-2, -1, 0, 1, 2, 3$
3.

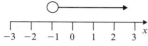

$$\begin{array}{ccccccc} -3 & -2 & -1 & 0 & 1 & 2 & 3 \end{array} \quad x$$

4. $x < 3$
5. Yes; they are both true for $-1 \leqslant x < 2$, for example, when $x = 1$

42. Expression, equation, identity or formula?

1. (a) expression   (b) equation   (c) equation
    (d) formula   (e) expression   (f) identity
2. For example: $a = 5, b = 2$: LHS $= 21 =$ RHS
              $a = 8, b = 3$: LHS $= 55 =$ RHS
              $a = -7, b = 10$: LHS $= -51 =$ RHS

43. Formulae

1. (a) £27   (b) 6 hours
2. $x = \dfrac{d - y}{2}$

44. Writing formulae

1. (a) $W = 7h + t$ (b) £47.50   (c) £11.25   (d) 12 hours
2. (a) $P = 15s$ (b) 150 cm   (c) 5 cm

45. Coordinates and midpoints

1. (a) (i) $(4, 3)$ (ii) $(4, -3)$
    (b) $(1, 0)$ or any point with y-coordinate 3 or -3
2. $(3, 3)$

46. Gradient

(a) $\frac{1}{2}$ (b) -1

47. $y = mx + c$

A: $y = x$ B: $y = -x + 2$ C: $x = 4$

48. Straight-line graphs

(a)

x	0	1	2	3
y	-1	1	3	5

(b)

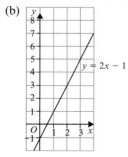

(c) gradient $= 2$

49. Formulae from graphs and tables

(a) $C = 50 + 25h$
(b)

Number of hours, h	0	1	2	3	4
Charges P (£)	50	75	100	125	150

(c)

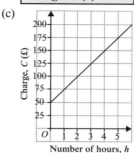

Number of hours, h

(d) 5 hours

50. Plotting quadratic graphs

(a)

x	-2	-1	0	1	2
y	3	0	-1	0	3

(b)

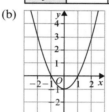

51. Real-life graphs

(a) approximately £35
(b) approximately €97–98
(c) approximately €130

52. Algebra problem-solving

(a) approximately 280 (275–285) g
(b) approximately 4.5 (4–5) oz
(c) approximately 2 lb 4 oz
(d) approximately 720 (710–730) g
(e) The conversion using the graph is not exact.

RATIO & PROPORTION

53. Metric measures

1. (a) 3 m   (b) 8 litres   (c) 7000 g
2. $2000 \div 200 = 10$

54. Time

1. 324 minutes
2. (a) $3\frac{5}{6}$ hours   (b) $3.8\dot{3}$ hours
3. 5 hours 45 minutes
4. 1.5 days is longest. 35 hours $\times$ 60 = 2100 minutes;
                1.5 days $\times$ 24 $\times$ 60 = 2160 minutes

55. Speed, distance, time

1. 20 seconds
2. 08:40 or 8.40 am

56. Percentage change

1. 92 340
2. 18% tax on £180 is £212.40 which is better value than 15% discount on £250 (£212.50)

57. Ratios

1. 4 : 9
2. £300

58. Proportion

1. $\frac{2}{5}$
2. Fastest: D; same rate: A and C; slowest: B

59. Direct proportion

1. (a) £6　　　(b) 50p　　　(c) £3.50
2. (a) 6048kr　　(b) £8.93

60. Inverse proportion

1.

Speed (km/h)	Time (h)	Speed × Time
15	4	60 km
12	**5**	60 km
24	2.5	60 km

2. (a) 16 hours　　(b) 4 hours

61. Distance–time graphs

(a)　10 minutes
(b)　1 hour 35 minutes
(c)　Horizontal from 19:30 to 19:40, slope from 19:40 to 19:50

62. Maps and scales

(a)　250 m　　(b) Yes; the distance is 200 m　　(c) 8 cm

63. Proportion problem-solving

1. Copper has a density of 9 g/cm³ which is more than the density of iron, 8 g/cm³
2. Dee; Mark's speed is 62.1 mph (450 ÷ 7.25) and Dee's is 62.5 mph (100 ÷ 1.6); Dee is travelling faster so will arrive first.

GEOMETRY & MEASURES

64. 3D shapes

1. (a) (i)　square-based pyramid　　(ii) cuboid
　　　　(iii) triangular prism
　　(b) tetrahedron
　　(c) cuboid
　　(d) triangle
2. C

65. Volume

1. 12 cm³
2. 94 cm³

66. Plans and elevations

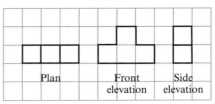

67. Measuring and drawing angles

1.

2. (a) 75°; acute angle　　(b) 145°; obtuse angle
3. 65°; the angle is an acute angle so it is less than 90°

68. Angles 1

$2x = 60°$

69. Angles 2

$a = 35°$ because alternate angles are equal.
$b = 35°$ because vertically opposite angles are equal.
$c = 50°$ because vertically opposite angles are equal.
$d = 50°$ because corresponding angles are equal.

70. Angles in polygons

1. 1800°
2. (a) 40°　　(b) 140°

71. Pythagoras' theorem

8.5 cm (1 d.p.)

72. Drawing triangles

1. Accurate drawing of triangle
2. Accurate construction of triangle

73. Constructing perpendicular lines

1. Accurate construction of perpendicular bisector
2.

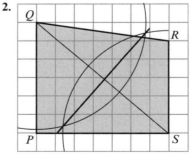

74. Constructing angles

1. Accurate construction of angle bisector
2. Accurate construction of 45° angle

75. Perimeter and area

1. (a) 22 cm　　(b) 22 cm²
2. (a) drawing of a shape with an area of 20 cm²
　　(b) correct perimeter of shape in part (a)
3. 30 cm

76. Area of rectangles and triangles

1. Two rectangles with dimensions 1 cm by 12 cm, 2 cm by 6 cm, 3 cm by 4 cm, or other sides that multiply to make 12.
2. 16 cm²
3. (a) 4 cm   (b) 10 cm²

77. Area of parallelograms and trapeziums

1. 42 cm²
2. 5 cm

78. Compound shapes

(a) $h = 4$ cm
(b) Area of triangle $= \frac{1}{2} \times 4 \times 10 = 20$ cm²
   Area of rectangle $= 8 \times 10 = 80$ cm²
   Area of shape $= 20 + 80 = 100$ cm²

79. Circumference

1. (a) 18.8 cm (1 d.p.)
   (b) 22.0 cm (1 d.p.)
2. 3.2 cm (1 d.p.)

80. Area of circles

1. (a) 1256.6 cm² (1 d.p.)
   (b) 19.6 m² (1 d.p.)
2. 13.7 cm² (1 d.p.)

81. Circles problem-solving

(a) 30.8 cm (1 d.p.)
(b) 56.5 cm² (1 d.p.)

82. Reflection

1.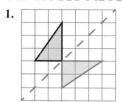

2. Reflection in the x-axis

83. Rotation

1.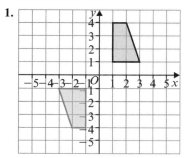

2. Rotation 90° clockwise about (0, 1)

84. Translation

1. Translation by vector $\begin{pmatrix} -3 \\ -1 \end{pmatrix}$

2.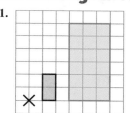

85. Enlargement

1.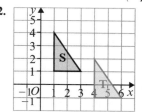

2. Enlargement, scale factor 3, centre of enlargement (0, 3)

86. Congruent and similar shapes

(a) A and C
(b) B and F

87. Shape problem-solving

75 cm²

PROBABILITY

88. Probability

(a) B
(b) $P(A) = \frac{2}{6} = \frac{1}{3}$

89. Outcomes

1. 0.2
2. 70%
3. (a) H1, H2, H3, T1, T2, T3
   (b) $\frac{1}{6}$

90. Experimental probability

1. (a) $\frac{14}{50}$
   (b) 56
2. 100

91. Probability diagrams

(a)

		First dice				
		1	2	3	4	5	6
Second dice	1	2	3	4	5	6	7
	2	3	4	5	6	7	8
	3	4	5	6	7	8	9
	4	5	6	7	8	9	10
	5	6	7	8	9	10	11
	6	7	8	9	10	11	12

(b) 36
(c) 7
(d) $\frac{3}{36} = \frac{1}{12}$

92. Probability tree diagrams

(a) $\frac{1}{2}$

(b) $\frac{1}{4}$

(c)

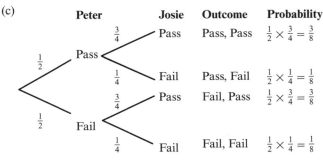

	Peter	Josie	Outcome	Probability
		Pass	Pass, Pass	$\frac{1}{2} \times \frac{3}{4} = \frac{3}{8}$
	Pass	Fail	Pass, Fail	$\frac{1}{2} \times \frac{1}{4} = \frac{1}{8}$
	Fail	Pass	Fail, Pass	$\frac{1}{2} \times \frac{3}{4} = \frac{3}{8}$
		Fail	Fail, Fail	$\frac{1}{2} \times \frac{1}{4} = \frac{1}{8}$

(d) P(Pass, Pass) $= \frac{1}{2} \times \frac{3}{4} = \frac{3}{8}$

93. Mutually exclusive and independent events

(a) Mutually exclusive; 5 is not even

(b) Mutually exclusive; 3 and 6 are multiples of 3 but are not smaller than 2

(c) Not mutually exclusive; 1 is a square number and an odd number

STATISTICS

94. Averages and range

2, 4, 6

95. Averages from tables

(a)

Height, h (cm)	Frequency	Midpoint, x	Frequency × midpoint ($f \times x$)
$0 < h \leqslant 20$	2	10	$10 \times 2 = 20$
$20 < h \leqslant 40$	3	30	$30 \times 3 = 90$
$40 < h \leqslant 60$	5	50	$50 \times 5 = 250$
$60 < h \leqslant 80$	2	70	$70 \times 2 = 140$
	Total frequency = 12		Total $f \times x = 500$

(b) Mean = 500 ÷ 12 = 41.7 cm (1 d.p.)

96. Stem and leaf diagrams

(a)

2	1 2 3 3 3
3	3 4 4 4 4 5 7 8
4	1 2 3

Key: 2|1 = 21 minutes

(b) 43 − 21 = 22 minutes

(c) 34 minutes

(d) 34 minutes

97. Analysing data

(a) 50

(b) A is not a good sample because it includes only people in her year, and older and younger students might enjoy different sports.

B is a good sample.

C is a biased sample because students are most likely to choose the sport they play.

D is a biased sample because boys and girls may prefer different sports.

98. Pie charts

(a) 36

(b) 10°

(c)

Members	Frequency	Angle
Boys	15	150°
Girls	10	100°
Adults	11	110°

(d) Chess club members

99. Scatter graphs

(a) As the temperature increases, the rainfall decreases.

(b) Negative correlation

(c) Line of best fit drawn

(d) 5–6 mm

100. Writing a report

(a) Soil A: mean 21.3 cm; median 25 cm; mode 25 cm; range 31 cm

Soil B: mean 18.5 cm; median 20 cm; mode 10 cm and 20 cm; range 36 cm

(b) Soil A

0	4 5
1	5 7
2	5 5 5
3	0 2 5

Soil B

0	2
1	0 0 3
2	0 0 3 4 5
3	8

Key: 2|1 = 21 minutes

(c) Comments such as:

Soil A is better for growing plants because it has a higher median and mean and lower range than Soil B.

The scientist could improve her investigation by: growing more plants; giving them more time to grow; changing the temperature.

Notes

Notes

Notes

Published by Pearson Education Limited, 80 Strand, London, WC2R 0RL.

www.pearsonschoolsandfecolleges.co.uk

Text and illustrations © Pearson Education Limited 2016
Copyedited by Charlotte Kelchner
Produced, typeset and illustrations by Cambridge Publishing Management Ltd
Cover illustration by Miriam Sturdee

The rights of Sharon Bolger and Bobbie Johns to be identified as authors of this work have been asserted by them in accordance
with the Copyright, Designs and Patents Act 1988.

First published 2016

18 17 16

10 9 8 7 6 5 4 3 2 1

British Library Cataloguing in Publication Data
A catalogue record for this book is available from the British Library

ISBN 9781292111544

Printed by L.E.G.O.